Lecture Notes in Chemistry

W0042958

Edited by:

Prof. Dr. Gaston Berthier
Université de Paris

Prof. Dr. Hanns Fischer
Universität Zürich

Prof. Dr. Kenichi Fukui
Kyoto University

Prof. Dr. George G. Hall
University of Nottingham

Prof. Dr. Jürgen Hinze
Universität Bielefeld

Prof. Dr. Joshua Jortner
Tel-Aviv University

Prof. Dr. Werner Kutzelnigg
Universität Bochum

Prof. Dr. Klaus Ruedenberg
Iowa State University

Prof Dr. Jacopo Tomasi
Università di Pisa

Springer
Berlin
Heidelberg
New York
Barcelona
Hong Kong
London
Milan
Paris
Singapore
Tokyo

S. Fraga   J.M. García de la Vega   E.S. Fraga

The Schrödinger
and Riccati Equations

 Springer

Authors

Serafín Fraga
Department of Chemistry
University of Alberta
Edmonton, AB, Canada T6G 2G2
and
Departamento de Química Física Aplicada
Universidad Autónoma de Madrid
E-28049 Canto Blanco (Madrid), Spain

Eric S. Fraga
Department of Chemical Engineering
University College London
Torrington Place
London WC1E 7JE, United Kingdom

José Manuel García de la Vega
Departamento de Química Física Aplicada
Universidad Autónoma de Madrid
E-28049 Canto Blanco (Madrid), Spain

Cataloging-in-Publication Data applied for

Die Deutsche Bibliothek - CIP-Einheitsaufnahme

Fraga, Serafin:
The Schrödinger and Riccati equations / S. Fraga ; J. M. García de la
Vega ; E. S. Fraga. - Berlin ; Heidelberg ; New York ; Barcelona ;
Hong Kong ; London ; Milan ; Paris ; Singapore ; Tokyo : Springer,
1999
  (Lecture notes in chemistry ; 70)

ISSN 0342-4901

ISBN 978-3-540-65105-5   ISBN 978-3-642-51458-6 (eBook)
DOI 10.1007/978-3-642-51458-6

Typesetting: Camera ready by authors
SPIN: 10691722   51/3143 - 543210 - Printed on acid-free paper

Preface

> While it can never be legitimately said of a
> theory that it is true, it can hopefully be said
> that it is the best available, that it is better
> than anything that has come before.
>
> A.F. Chalmers

The Riccati equation has represented for a long time a mere mathematical curiosity but the transcendence of the Schrödinger equation was apparent from the moment of its inception.

Its mathematical significance as a differential equation, the difficulties to be faced in its solution, and the relevance of the information to be obtained have encouraged the development of countless formulations, methods, techniques, and calculations resulting in a wealth of results and interpretations of physical and chemical phenomena. Technological advances have played a decisive role in the progress in this field and the availability of fast, large-capacity computers makes it possible to envisage the possibility of more and more accurate results.

Traditional approaches, such as those originating from the perturbation method and the variational principle, have been discussed exhaustively in the literature and consequently they will not be included in this work. Rather, our attention will be focused on those aspects which, for a variety of reasons, have not attracted so much attention. Circumstances change with time and such topics as Schrödinger local energies, the Schrödinger equation in momentum space, and the Riccati equation are now worthwhile of a renewed interest.

It only remains to gratefully acknowledge the dedicated collaboration of J. Jorgensen, who has overcome as usual all the technical problems in the preparation of this work.

S. Fraga   J.M. García de la Vega   E.S. Fraga
Edmonton   Madrid   London

Table of Contents

Acknowledgments

Theoretical developments and formulations from many sources have been summarized or adapted for use in this work. We would like to acknowledge here our indebtedness to those organizations from whose publications information has been taken and/or copyright material has been used:

> Academic Press, Addison-Wesley, American Institute of Physics, American Physical Society, Benjamin, Benjamin-Cummings, Butterworth, Cambridge University Press, Chapman and Hall, Dover, Elsevier, European Physical Society, Herman et Cie., Holt Rinehart and Winston, Institute of Physics Publishing, Interscience, John Wiley and Sons, Kluwer, Les Editions de Physique, Longman, MacMillan, Manchester University Press, Masson et Cie., McGraw-Hill, National Research Council (Canada), North-Holland, Oxford University Press, Pergamon, Physical Society (Japan), Plenum, Prentice-Hall, Princeton University Press, Reidel, Royal Society (London), Royal Swedish Academy of Sciences, Schaum, Springer, Van Nostrand Reinhold, Wiley-VCH, and World Scientific.

Specific mention must be made, in particular, of the following papers from which copyright material has been used, and of the publishers of the journal in which those papers appeared:

ACADEMIC PRESS, INC.
Copyright © 1959 Academic Press, Inc., Orlando, Florida, U.S.A. Adapted by permission of Academic Press, Inc.

> P.O. Löwdin. *Scaling problem, virial theorem, and connected relations in Quantum Mechanics*. Journal of Molecular Spectroscopy *3*, 46 (1959).

AMERICAN INSTITUTE OF PHYSICS
Copyright © 1942, 1963, 1966, 1966, 1976, 1995 American Institute of Physics, Woodbury, NY, U.S.A. Adapted by permission of the American Institute of Physics.

> A.A. Frost. *The approximate solution of the Schrödinger equations by a least squares method*. Journal of Chemical Physics *10*, 240 (1942).

S. Fraga wishes to acknowledge the continued research support received from the National Research Council of Canada and from the Natural Sciences and Engineering Research Council of Canada during the past thirty-five years.

J.M. Garcia de la Vega gratefully acknowledges the financial support of the Ministerio de Educación y Cultura (Spain) during his sabbatical leave at the Department of Chemistry of the University of Alberta, Edmonton, Alberta, Canada.

E.S. Fraga gratefully acknowledges the support provided by an Advanced Fellowship from the United Kingdom's Engineering and Physical Sciences Research Council (*EPSRC*).

1 Introduction

The Schrödinger and Riccati equations are the topics to be discussed in this work, together with the Schrödinger-Riccati equation, thus denoted in order to reflect the fact that it has been derived from the other two.

The Riccati equation

$$y' = p(x) + q(x)y + r(x)y^2 \tag{1.1}$$

represents an extension of the simple first-order, linear ordinary differential equation. Although proposed about three centuries ago (Simmons 1972), it has not attracted particular attention until recently, when it started appearing in quantum treatments.

The time-independent linear Schrödinger equation, on the other hand, may be obtained (plus relativistic corrections and magnetic terms) in the reduction to non-relativistic form of the extended Dirac-Breit equation. Within that context, the Schrödinger equation may be seen as just one component, albeit a very important one, of the complete description given by relativistic quantum mechanics.

In spite of this formal deficiency, the Schrödinger equation has evolved into the fundamental tool for quantum treatments and analogies have been proposed in order to show how it may be considered a step in the evolution, through the Lagrange and Hamilton equations, from the Newtonian laws of motion.

The analogy was established with the wave equation for the propagation of light waves as well as for vibratory phenomena (such as vibrating strings and membranes) but in this connection it is worthwhile to note the difficulties in obtaining a satisfactory derivation of the differential equation describing a physical phenomenon (Simmons 1972).

The wave equation for the propagation of light (Iñiguez 1949, Kemble 1958, Zatzkis 1960a) is

$$v^2 \, \nabla^2 \Phi = \frac{\partial^2 \Phi}{\partial t^2} \tag{1.2}$$

where Φ gives at each point the amplitude of the vibration and v is the velocity (local phase velocity, velocity of transmission, velocity of the disturbance). The velocity may depend on the frequency ν (and therefore on the position coordinates) and on the time (whenever the medium is in motion). ∇^2 is the Laplacian operator

$$\nabla^2 = \nabla \cdot \nabla = \frac{\partial^2}{\partial x^2} + \frac{\partial^2}{\partial y^2} + \frac{\partial^2}{\partial z^2} \tag{1.3}$$

related to the gradient operator

$$\nabla = \mathbf{i}\frac{\partial}{\partial x} + \mathbf{j}\frac{\partial}{\partial y} + \mathbf{k}\frac{\partial}{\partial z} \tag{1.4}$$

One may now assume that matter waves may be described by an equation similar to Eq. (1.2). Taking into account the fundamental relationships

$$E = h\nu \qquad\qquad p = h/\lambda$$

(relating the energy E, the linear momentum p, the wave length λ, and the frequency ν, and where h is the Planck constant) as well as the basic relationships

$$E = T + V \qquad\qquad T = \frac{1}{2}mv^2 \qquad\qquad p = mv$$

(where T and V denote the kinetic and potential energy, respectively, p is the linear momentum, and m is the mass) one obtains

$$v = \lambda\nu = E/p = E[2m(E - V)]^{-\frac{1}{2}}$$

Substitution into Eq. (1.2) yields

$$\frac{E^2}{2m(E - V)}\nabla^2\Phi = \frac{\partial^2\Phi}{\partial t^2} \tag{1.5}$$

If the potential energy V is independent of t, a separation of variables

$$\Phi(x,y,z,t) = \Psi(x,y,z)f(t) \tag{1.6}$$

is possible and one can write

$$\frac{E^2}{2m(E - V)\Psi}\nabla^2\Psi = \frac{1}{f}\frac{d^2f}{dt^2} \tag{1.7}$$

Because of the independence of the variables, the two sides of this equation must be equal to a constant, say k. One obtains then

$$\frac{E^2}{2m(E-V)\Psi} \, \nabla^2\Psi = k \tag{1.8a}$$

$$\frac{1}{f} \frac{d^2f}{dt^2} = k \tag{1.8b}$$

On the basis of physical considerations (Kemble 1958) it may be justified that

$$k = -4\pi^2\nu^2 = -\frac{4\pi^2 E^2}{h^2} = -\frac{E^2}{\hbar^2}$$

(with $\hbar = h/2\pi$), so that Eqs. (1.8) become

$$-(\hbar^2/2m)\nabla^2\Psi + V\Psi = E\Psi \tag{1.9a}$$

$$\frac{d^2f}{dt^2} = -4\pi^2\nu^2 f \tag{1.9b}$$

Equation (1.9a) is the time-independent linear Schrödinger equation for a single particle, with mass m, moving in the presence of a potential V. It is because of this analogy that the solutions Ψ of the Schrödinger equation were denoted as wave functions, a designation that still perdures.

Quite independently, the Schrödinger equation may be seen (Pauling and Wilson 1935, Davydov 1965) as an evolution of the equation

$$H(x,y,z,p_x,p_y,p_z) = T(p_x,p_y,p_z) + V(x,y,z) = E \tag{1.10}$$

of classical Newtonian mechanics, where H is the Hamiltonian function and p_x, p_y, p_z are the components of the linear momentum. If the Hamiltonian function does not contain products of position coordinates and momenta or contains such products which will involve only commuting operators (after the change to operator form) (Davydov 1965), one may propose the formal transformations

$$p_x \rightarrow \frac{\hbar}{i} \frac{\partial}{\partial x} \quad p_y \rightarrow \frac{\hbar}{i} \frac{\partial}{\partial y} \quad p_z \rightarrow \frac{\hbar}{i} \frac{\partial}{\partial z} \quad E \rightarrow -\frac{\hbar}{i} \frac{\partial}{\partial t}$$

and obtain, for a particle with mass m, the equation

$$-\frac{\hbar^2}{2m}\left(\frac{\partial^2\Phi}{\partial x^2} + \frac{\partial^2\Phi}{\partial y^2} + \frac{\partial^2\Phi}{\partial z^2}\right) + V\Phi = -\frac{\hbar}{i} \frac{\partial\Phi}{\partial t} \tag{1.11}$$

which is the time-dependent linear Schrödinger equation for a single particle. Proceeding again with a separation of variables

$$\Phi(x,y,z,t) = \Psi(x,y,z)f(t)$$

one obtains

$$-\frac{\hbar^2}{2m\Psi} \nabla^2\Psi + V = -\frac{\hbar}{if}\frac{df}{dt} \qquad (1.12)$$

With the same argument as before, regarding the equality of both sides to a constant, and adopting E as such a constant we obtain

$$-\frac{\hbar^2}{2m} \nabla^2\Psi + V\Psi = E\Psi \qquad (1.13a)$$

$$\frac{df}{dt} = -\frac{i}{\hbar} Ef \qquad (1.13b)$$

where again Eq. (1.13a) is the time-independent linear Schrödinger equation for a single particle, with mass m, moving in the presence of a potential V. In the case of a many-particle system, the corresponding Schrödinger equation will be written as

$$-\frac{\hbar^2}{2} \sum_i \frac{1}{m_i} \nabla_i^2\Psi + V\Psi = E\Psi \qquad (1.14)$$

where the summation extends to all the particles in the system, and m_i and ∇_i^2 represent the mass and the Laplacian operator for particle i. In short, as customary, Eq. (1.14) may be rewritten as

$$\mathcal{H}\Psi = E\Psi \qquad (1.15)$$

where

$$\mathcal{H} = -\frac{\hbar^2}{2} \sum_i \frac{1}{m_i} \nabla_i^2 + V \qquad (1.16)$$

is the Hamiltonian operator. There is no need of further discussion at this point, as Eq. (1.15) will be treated in detail in the next chapters. Here it must be emphasized, however, that *the above development does not constitute a derivation of the Schrödinger equation.*

The Schrödinger-Riccati equation may be developed by joint consideration of both the Schrödinger and the Riccati equations. Succinctly, the development may be summarized as follows. Assuming that an approximation ϕ to the exact solution Ψ is known, one can write

$$\Psi = \phi + \varphi$$

where φ denotes the correction to be determined. Equation (1.15) then becomes

$$\mathcal{H}\Psi = \mathcal{H}\phi + \mathcal{H}\varphi = E\phi + E\varphi = E\Psi$$

which may be rewritten as

$$(\mathcal{H} - E)\phi = - (\mathcal{H} - E)\varphi = S^{(0)}$$

Defining the quantities

$$\varepsilon_\phi = \frac{\mathcal{H}\phi}{\phi} \qquad \varepsilon_\varphi = \frac{\mathcal{H}\varphi}{\varphi}$$

it can be proven that

$$\Psi = \frac{\varepsilon_\phi - \varepsilon_\varphi}{E - \varepsilon_\varphi} \phi = \frac{\varepsilon_\varphi - \varepsilon_\phi}{E - \varepsilon_\phi} \varphi$$

may be expressed in terms of ε_ϕ and ε_φ, for which one can write the Riccati equation

$$\varepsilon^{(1)} = \frac{1}{S^{(0)}} (E - \varepsilon)(\varepsilon - H^{(1)})$$

where ε stands for either ε_ϕ or ε_φ and with

$$H^{(1)} = \frac{\partial \mathcal{H}\phi}{\partial \phi}$$

After some manipulation, presented in detail in the corresponding chapter below, one arrives at the algebraic Schrödinger-Riccati equation

$$\sum_{n=0}^{\infty} \frac{1}{n!} \varphi^n S^{(n)} = 0 \qquad\qquad (1.17)$$

where

$$S^{(1)} = H^{(1)} - E \qquad S^{(n)} = \frac{\partial^n \mathcal{H}\phi}{\partial \phi^n} \qquad (\text{for } n > 1)$$

The solution of Eq. (1.17), to the appropriate degree, will yield φ (for known E).

The following sections will present, to the extent possible in each case, the derivation/formal justification, solution, characteristics, and applications of the Riccati, time-dependent and time-independent linear Schrödinger, and Schrödinger-Riccati equations.

The non-linear Schrödinger equation

$$i\frac{\partial \Phi}{\partial t} + \nabla^2\Phi + q\left|\Phi\right|^2\Phi = 0 \qquad\qquad (1.18)$$

where $\Phi(x,y,z,t)$ is a complex-valued function in the domain $-\infty<x,y,z<\infty$, $t>0$, and q is a real-valued parameter, is a non-linear, dispersive wave equation. As such it lies outside of the context of quantum mechanics but it will also be discussed, albeit briefly, because it transforms into a linear Schrödinger equation for q = 0. This equation describes the evolution of any weakly non-linear, strongly dispersive almost monochromatic wave, exhibiting also soliton solutions for appropriate initial conditions. This equation appears in the formulation of phenomena in such diverse fields as superconductivity, plasma/low temperature/condensed matter physics, deep water waves and, particularly, non-linear optics. Examples of the latter case are the propagation of a laser beam through plasma, the behaviour of a plane stationary beam in a medium with a non-linear refractive index, a quasi-monochromatic one-dimensional wave in a medium with dispersion and inertialess non-linearity, the dynamics of a non-linear short optical pulse envelope in a fiber, etc. [See, e.g., the work of Kelley (1965), Talanov (1965), Zhakarov (1966), Bespalov *et al.* (1968), Zhakarov and Shabat (1972), Hasegawa (1989), Taylor (1992), Abdullaev *et al.* (1993), Radhakrishnan and Lakshmanan (1996), as well as the additional references given in Chapter 7.] It is also interesting to note that the Riccati equation appears in the formalism of soliton theory (Fordy 1990).

The Linear Schrödinger Equation

In this section we will first establish the status of the time-independent linear Schrödinger equation within the framework of relativistic quantum mechanics.

The Dirac equation for the electron and the two-particle Breit correction constitute the starting points for the development of the Dirac-Breit equation for a many-particle system.

The generalized Dirac-Breit equation is first transformed with separation of the coordinates of the centre of mass and then reduced to non-relativistic form. For atoms, a final transformation of the non-relativistic equation, with elimination of the motion of the centre of mass, yields the desired Schrödinger equation plus the specific mass correction. For molecules, however, the Schrödinger equation normally used in computations must be justified on the basis of the Born-Oppenheimer approximation.

The Schrödinger equation is usually expressed in position space but it is also possible to express it, through a Fourier transform, in momentum space and both forms will be discussed. The existence of solutions and their characteristics will be examined and particular attention will be given to the Hellmann-Feynman, hypervirial, and virial theorems and the subject of local energies as well as to the related null-kinetic and null-potential energy regions.

This section will be concluded with an examination of the time-dependent Schrödinger equation.

2 Derivation of the Schrödinger Equation

The goal of this Chapter is to examine the appearance of the Schrödinger equation in the reduction of the generalized Dirac-Breit equation for a many-particle system to non-relativistic form. As our interest is focused on the Schrödinger equation we will not enter into a detailed discussion of relativistic quantum mechanics, restricting ourselves to the essential points needed for the development of the formulation.

2.1 The Dirac Equation

The Dirac equation for the electron (Dirac 1928ab, 1958), using the notation of Bethe and Salpeter (1957), may be written as

$$\left\{ \frac{i\hbar}{c} \frac{\partial}{\partial t} - \boldsymbol{\alpha} \cdot \mathbf{p} - \beta mc \right\} \Phi(\mathbf{r},t) = 0 \tag{2.1}$$

where c is the speed of light, m is the mass of the electron, $\mathbf{p}$ is the linear momentum vector operator ($\mathbf{p} = -i\hbar \nabla$), and $\hbar = h/2\pi$ (where h is Planck constant). β is a matrix of order 4

$$\beta = \begin{pmatrix} \mathbf{I} & \mathbf{O} \\ \mathbf{O} & -\mathbf{I} \end{pmatrix}$$

where $\mathbf{O}$ is the null matrix of order 2 and $\mathbf{I}$ is the unit matrix

$$\mathbf{I} = \begin{pmatrix} 1 & 0 \\ 0 & 1 \end{pmatrix}$$

of order 2. The Cartesian coordinates of the vector operator $\boldsymbol{\alpha}$ are the matrices of order 4

$$\alpha_x = \begin{pmatrix} 0 & \sigma_x^P \\ \sigma_x^P & 0 \end{pmatrix} \qquad \alpha_y = \begin{pmatrix} 0 & \sigma_y^P \\ \sigma_y^P & 0 \end{pmatrix} \qquad \alpha_z = \begin{pmatrix} 0 & \sigma_z^P \\ \sigma_z^P & 0 \end{pmatrix}$$

where

$$\sigma_x^P = \begin{pmatrix} 0 & 1 \\ 1 & 0 \end{pmatrix} \qquad \sigma_y^P = \begin{pmatrix} 0 & -i \\ i & 0 \end{pmatrix} \qquad \sigma_z^P = \begin{pmatrix} 1 & 0 \\ 0 & -1 \end{pmatrix}$$

are the Pauli matrices of order 2 (Pauli 1927). The function Φ is a column vector of order 4, whose elements are functions of the position of the electron. The Dirac equation is fully invariant under a Lorentz transformation. [See, e.g., the review of Zatzkis (1960b) for details on the Lorentz transformation.]

In order to take into account the existence of an external electromagnetic field, use is made of the transformation

$$\frac{i\hbar}{c}\frac{\partial}{\partial t} \to \frac{i\hbar}{c}\frac{\partial}{\partial t} + \frac{e}{c}A_0(\mathbf{r}) \qquad \mathbf{p} \to \mathbf{p} + \frac{e}{c}\mathbf{A}(\mathbf{r})$$

where $A_0(\mathbf{r})$ and $\mathbf{A}(\mathbf{r})$ are the scalar and vector potentials of the external electromagnetic field at position $\mathbf{r}$ and e is the charge of the electron (in absolute value). Equation (2.1) then becomes

$$\mathcal{H}_D\Phi(\mathbf{r},t) = i\hbar\frac{\partial\Phi(\mathbf{r},t)}{\partial t} \tag{2.2}$$

where

$$\mathcal{H}_D = \beta\, mc^2 - e\, A_0(\mathbf{r}) + \alpha \cdot (c\mathbf{p} + e\mathbf{A}(\mathbf{r})) \tag{2.3}$$

is the Dirac Hamiltonian operator. For stationary states, using the variable separation

$$\Phi(\mathbf{r},t) = \Psi(\mathbf{r})e^{-iEt/\hbar}$$

one obtains

$$\mathcal{H}_D\Psi(\mathbf{r}) = E\Psi(\mathbf{r}) \tag{2.4}$$

where E is the energy of the system.

The scalar potential $A_0(\mathbf{r})$ may contain a nuclear Coulomb potential. In the particular case when $A_0(\mathbf{r}) = -Ze^2/r$ (where Z is the nuclear charge, in e-units) and $\mathbf{A}(\mathbf{r}) = 0$, Eq. (2.2) will describe the motion of the electron in a Coulomb field for the discrete spectrum (Darwin 1928, Gordon 1928).

2.2 The Breit Correction

The Dirac-Breit equation for two electrons, including the Breit correction (Breit 1929, 1930, 1932), may be written for a stationary state as

$$\mathcal{H}_{DB}\Psi(\mathbf{r}_1,\mathbf{r}_2) = E\Psi(\mathbf{r}_1,\mathbf{r}_2) \tag{2.5}$$

with

$$\mathcal{H}_{DB} = \mathcal{H}_1 + \mathcal{H}_2 + \frac{e^2}{r_{12}} + \mathcal{H}_B$$

where

$$\mathcal{H}_i = \beta_i mc^2 - eA_0(\mathbf{r}_i) + \alpha_i \bullet (c\mathbf{p}_i + eA(\mathbf{r}_i)) \tag{2.6}$$

is the Dirac Hamiltonian operator for electron i, $A_0(\mathbf{r}_i)$ and $A(\mathbf{r}_i)$ being the scalar (including the nuclear Coulomb term) and vector potentials of the external electromagnetic field acting on the electron at position $\mathbf{r}_i$. The term e^2/r_{12} is the electron repulsion energy and the Breit correction is given by

$$\mathcal{H}_B = - \frac{e^2}{r_{12}}\left[\alpha_1 \bullet \alpha_1 + \frac{(\alpha_1 \bullet \mathbf{r}_{12})(\alpha_2 \bullet \mathbf{r}_{12})}{r_{12}^2}\right] \tag{2.7}$$

r_{12} being the interelectronic separation. The operators α_i and β_i act on the components of Ψ for electron i.

The Dirac-Breit equation is not fully Lorentz invariant but it is appropriate for the purpose in mind in spite of its deficiencies (Grotch and Hegstrom 1971, Hegstrom 1973).

2.3 The Generalized Dirac-Breit Equation

The generalized Dirac-Breit equation for a stationary state of a system of particles will be written as

$$\mathcal{H}_{GDB}\Phi(\mathbf{r}_1,\mathbf{r}_2,...) = E\Phi(\mathbf{r}_1,\mathbf{r}_2,...) \tag{2.8}$$

with

$$\mathcal{H}_{GDB} = \sum_1 \mathcal{H}_i + \sum_{i<j} \mathcal{U}_{ij} \tag{2.9}$$

and where

$$\mathcal{H}_i = \beta_i m_i c^2 + \alpha_i \cdot c\mathbf{p}_i \tag{2.10a}$$

$$u_{ij} = \frac{e_i e_j}{r_{ij}} \left[1 - \frac{(\alpha_i \cdot \alpha_j)}{2} - \frac{(\alpha_i \cdot \mathbf{r}_{ij})(\alpha_j \cdot \mathbf{r}_{ij})}{2r_{ij}^2} \right] \tag{2.10b}$$

The particles are characterized by their masses m_i and charges e_i (with the appropriate sign), r_{ij} denotes the inter-particle separation, and all the remaining symbols have the same meaning as above. The summations extend to all the particles in the system. The extended Hamiltonian of Hegstrom (1973) contains, in addition, the interactions with an external electromagnetic field as well as the appropriate internal magnetic interactions (with consideration of the anomalous magnetic moments of the particles). All those terms are omitted here because they are irrelevant for the purpose of this work (see below).

The transformation of the generalized Dirac-Breit equation to non-relativistic form is performed in two steps, as discussed below.

2.3.1 Separation of the Coordinates of the Centre of Mass

The first step in the transformation of Eq. (2.8) consists of the separation of the coordinates of the centre of mass, as carried out by Hegstrom (1973).

The position and momentum vectors of the centre of mass are given by

$$\mathbf{R} = \frac{1}{M} \sum_i m_i \mathbf{r}_i \tag{2.11a}$$

$$\mathbf{P} = \sum_i \mathbf{p}_i \tag{2.11b}$$

where the position vectors $\mathbf{r}_i$ of the particles are referred to the laboratory frame; M denotes the total mass of the system. The internal coordinates, referred to a given particle a (which may be a nucleus, but not necessarily), are defined by

$$\mathbf{r}_{ia} = \mathbf{r}_i - \mathbf{r}_a \tag{2.12}$$

and the relationships between $\mathbf{R}$, the coordinates in the laboratory frame, and the internal coordinates are

$$\mathbf{R} = \frac{m_a}{M}\mathbf{r}_a + \sum_{j \neq a} \frac{m_j}{M}\mathbf{r}_j = \mathbf{r}_a - \frac{M-m_a}{M}\mathbf{r}_a + \sum_{j \neq a} \frac{m_j}{M}\mathbf{r}_j$$

$$= \mathbf{r}_a - \frac{1}{M}\left(\sum_{j \neq a} m_j\right)\mathbf{r}_a + \sum_{j \neq a} \frac{m_j}{M}\mathbf{r}_j = \mathbf{r}_a + \sum_{j \neq a} \frac{m_j}{M}(\mathbf{r}_j - \mathbf{r}_a) \qquad (2.13a)$$

$$= \mathbf{r}_a + \sum_{j \neq a} \frac{m_j}{M}\mathbf{r}_{ja}$$

$$\mathbf{R} = \mathbf{r}_a + \sum_{j \neq a} \frac{m_j}{M}\mathbf{r}_{ja} = \mathbf{r}_i + \mathbf{r}_a - \mathbf{r}_i + \sum_{j \neq a} \frac{m_j}{M}\mathbf{r}_{ja}$$

$$= \mathbf{r}_i - \mathbf{r}_{ia} + \sum_{j \neq a} \frac{m_j}{M}\mathbf{r}_{ja} \qquad (\text{for } i \neq a) \qquad (2.13b)$$

The corresponding relationships for the conjugate momenta may be easily obtained as follows. Let us consider first the set of particles from which particle a is omitted and define the quantities

$$M_J = \sum_{j \neq a} m_j = M - m_a$$

$$R_J = \frac{1}{M_J} \sum_{j \neq a} m_j \mathbf{r}_j$$

$$P_J = \sum_{j \neq a} \mathbf{p}_j = \sum_{j \neq a} m_j \dot{\mathbf{r}}_j$$

where the notation $\dot{\mathbf{r}}_j$ is used to denote differentiation with respect to t (as is customary). The system may now be treated as a two-particle problem, with masses m_a and M_J at positions $\mathbf{r}_a$ and $\mathbf{R}_J$, respectively. One can then write (see, e.g., Noggle 1985)

$$\dot{\mathbf{r}}_a = \dot{\mathbf{R}} - \frac{\mu}{m_a}\dot{\mathbf{R}}_{Ja} = \dot{\mathbf{R}} - \frac{\mu}{m_a}(\dot{\mathbf{R}}_J - \dot{\mathbf{r}}_a) \qquad (2.14)$$

where

$$\mu = \frac{m_a M_J}{m_a + M_J} = \frac{m_a M_J}{M}$$

is the reduced mass. Equation (2.14) may be rewritten as

$$\dot{\mathbf{r}}_a = \dot{\mathbf{R}} - \frac{1}{m_a}\left\{ \sum_{j \neq a} \mu \frac{m_j}{M_J} (\dot{\mathbf{r}}_j - \dot{\mathbf{r}}_a) \right\} = \dot{\mathbf{R}} - \frac{1}{m_a} \sum_{j \neq a} \mathbf{p}_{ja}$$

and finally (after multiplication by m_a) as

$$\mathbf{p}_a = \frac{m_a}{M}\mathbf{P} - \sum_{j \neq a} \mathbf{p}_{ja} \qquad (2.15a)$$

Taking into account Eq. (2.11b) one can write

$$\mathbf{p}_a = \mathbf{P} - \sum_{j \neq a} \mathbf{p}_j = \frac{m_a}{M}\mathbf{P} - \sum_{j \neq a} \mathbf{p}_{ja}$$

which can be rewritten as

$$\sum_{j \neq a} \left(\frac{m_j}{M}\mathbf{P} - \mathbf{p}_j + \mathbf{p}_{ja} \right) = 0$$

which is satisfied if

$$\mathbf{p}_i = \frac{m_i}{M}\mathbf{P} + \mathbf{p}_{ia} \qquad (2.15b)$$

for every particle i ($\neq a$).

The Hamiltonian operator $\mathcal{H}_{GDB}$ and $\mathbf{P}$ commute and therefore the function Ψ will be a simultaneous eigenfunction of both operators. The eigenequation for $\mathbf{P}$ is

$$\mathbf{P}\psi(\mathbf{R},\mathbf{r}_{1a},\mathbf{r}_{2a}, ...) = -i\hbar \nabla_{\mathbf{R}}\psi(\mathbf{R},\mathbf{r}_{1a},\mathbf{r}_{2a}, ...) = \mathbf{K}\psi(\mathbf{R},\mathbf{r}_{1a},\mathbf{r}_{2a}, ...)$$

where $\mathbf{K}$ is a constant vector, independent of $\mathbf{R}$, and its solution will be

$$\psi(\mathbf{R},\mathbf{r}_{1a},\mathbf{r}_{2a}, ...) = \Psi(\mathbf{r}_{1a},\mathbf{r}_{2a}, ...)e^{i\,\mathbf{K}\cdot\mathbf{R}} \qquad (2.16)$$

Taking into account Eqs. (2.15a) and (2.15b), one can see that the terms of $\mathcal{H}_i$, Eq. (2.10a), dependent on $\mathbf{p}_a$ and $\mathbf{p}_i$ (for all the other particles), when acting on ψ will yield

$$\mathbf{p_a} \, \psi \, (\mathbf{R}, \mathbf{r}_{1a}, \mathbf{r}_{2a}, \dots) = \left\{ \frac{m_a}{M} \mathbf{P} - \sum_{j \neq a} \mathbf{p}_{ja} \right\} \psi \, (\mathbf{R}, \mathbf{r}_{1a}, \mathbf{r}_{2a}, \dots)$$

$$= \left\{ \frac{m_a}{M} \mathbf{K} - \sum_{j \neq a} \mathbf{p}_{ja} \right\} \psi \, (\mathbf{R}, \mathbf{r}_{1a}, \mathbf{r}_{2a}, \dots)$$

$$\mathbf{p_i} \, \psi \, (\mathbf{R}, \mathbf{r}_{1a}, \mathbf{r}_{2a}, \dots) = \left\{ \frac{m_i}{M} \mathbf{P} + \mathbf{p}_{ia} \right\} \psi \, (\mathbf{R}, \mathbf{r}_{1a}, \mathbf{r}_{2a}, \dots)$$

$$= \left\{ \frac{m_i}{M} \mathbf{K} + \mathbf{p}_{ia} \right\} \psi \, (\mathbf{R}, \mathbf{r}_{1a}, \mathbf{r}_{2a}, \dots)$$

Therefore, taking into account Eq. (2.16), we can write

$$\mathcal{H}'_{GDB} \, \Psi(\mathbf{r}_{1a}, \mathbf{r}_{2a}, \dots) = E\Psi(\mathbf{r}_{1a}, \mathbf{r}_{2a}, \dots) \tag{2.17}$$

where $\mathcal{H}'_{GDB}$ is defined [see Eq. (2.9)] as

$$\mathcal{H}'_{GDB} = \sum_i \mathcal{H}'_i + \sum_{i<j} \mathcal{U}_{ij} \tag{2.18}$$

with $\mathcal{U}_{ij}$ given by Eq. (2.10b) and with the operators $\mathcal{H}'_i$ expressed as in Eq. (2.10a) but in terms of

$$\mathbf{P_a} = \frac{m_a}{M} \mathbf{K} - \sum_{j \neq a} \mathbf{p}_{ja} \tag{2.19a}$$

$$\mathbf{P_i} = \frac{m_i}{M} \mathbf{K} + \mathbf{p}_{ia} \qquad (i \neq a) \tag{2.19b}$$

That is: the transformed operator $\mathcal{H}'_{GDB}$ depends only on internal coordinates.

2.3.2 Reduction to Non-Relativistic Form

The transformation to a non-relativistic form may be achieved by a Foldy-Wouthuysen-type transformation (Foldy and Wouthuysen 1950, Foldy 1962), using the method of Chraplyvy (1953ab) and Barker and Glover (1955), as applied by Grotch and Hegstrom (1971) and Hegstrom (1973) to two- and many-particle systems, respectively. It must be mentioned, however, that the method of small components and the method of Foldy-Wouthmysen are identical, at least in the case of a free electron and that it is expected (Moss 1971) that this equivalence may hold also when the electron is in the presence of an electromagnetic field. [For details on the method of small components, see the work of Moss (1973) and Dyall (1997); see also the recent general review of Almlöf and Groppen (1996).]

The Chraplyvy-Barker-Glover unitary transformation of a two-particle operator

$$\mathcal{H}_{ij} = \beta_i m_i c^2 + \beta_j m_j c^2 + \mathcal{H}_{ij}^{ee} + \mathcal{H}_{ij}^{oe} + \mathcal{H}_{ij}^{eo} + \mathcal{H}_{ij}^{oo} \qquad (2.20a)$$

yields (Chraplyvy 1953ab) an expansion

$$\mathcal{H}_{ij} = \beta_i m_i c^2 + \beta_j m_j c^2 + \mathcal{H}_{ij}^{ee} + \frac{1}{2m_i c^2} \beta_i (\mathcal{H}_{ij}^{oe})^2 + \frac{1}{2m_j c^2} \beta_j (\mathcal{H}_{ij}^{eo})^2 + \dots \qquad (2.20b)$$

in terms of inverse powers of c and of the masses of the two particles. The superscripts e and o indicate whether the operators are even or odd with respect to the Dirac matrices for the two particles. The transformed Hamiltonian operator still yields a 16-component eigenequation, which is then reduced to an *approximate* 4-component eigenequation by elimination of twelve components. The resulting equation compares with the Breit equation reduced to the corresponding 4-component form obtained (Breit 1929) by the method of large components. [See, however, the observations of Chraplyvy (1953a) in this regard.]

In the present case, not being interested in the higher-order mass corrections, only the terms given explicitly in Eq. (2.20b) need be considered, which means that the starting Hamiltonian to be used is

$$\mathcal{H} = \sum_i \left\{ \beta_i m_i c^2 + \alpha_i \cdot c p_i + \sum_{j>i} \frac{e_i e_j}{r_{ij}} \right\}$$

$$= \beta_a m_a c^2 + \sum_{i \neq a} \beta_i m_i c^2 + \alpha_a \cdot c p_a + \sum_{i \neq a} \alpha_i \cdot c p_i + \sum_{i \neq a} \frac{e_i e_a}{r_{ia}} + \sum_{i \neq a} \sum_{j>i} \frac{e_i e_j}{r_{ij}}$$

and the final non-relativistic Hamiltonian operator is then (Bethe and Salpeter 1957, Hegstrom 1973)

$$\mathcal{H} = \sum_i \left\{ m_i c^2 + \frac{1}{2m_i} p_i \cdot p_i + \sum_{j>i} \frac{e_i e_j}{r_{ij}} \right\} = Mc^2 + \sum_i \left\{ \frac{1}{2m_i} p_i \cdot p_i + \sum_{j>i} \frac{e_i e_j}{r_{ij}} \right\}$$

2.4 The Schrödinger Hamiltonian Operator

The first term of the *rhs* of the above equation is constant and therefore for the study of the motion of the particles we will use the Schrödinger Hamiltonian operator

$$\mathcal{H} = \sum_i \left\{ \frac{1}{2m_i} \mathbf{p}_i \cdot \mathbf{p}_i + \sum_{j>i} \frac{e_i e_j}{r_{ij}} \right\} \tag{2.21}$$

One (in atoms) or more (in molecules) of those particles are nuclei which, being much heavier than the electrons, will move much more slowly. Consequently it seems appropriate to try and separate their motions. The separation of the motions is straightforward in the case of atoms but not so in the case of molecules, as discussed below.

2.4.1 Atoms

Taking into account Eqs. (2.19), Eq. (2.21) may be rewritten as

$$\mathcal{H} = \frac{1}{2m_a} \left(\frac{m_a}{M} \mathbf{K} - \sum_{i \neq a} \mathbf{p}_{ia} \right)^2 + \sum_{i \neq a} \frac{1}{2m_i} \left(\frac{m_i}{M} \mathbf{K} + \mathbf{p}_{ia} \right)^2 + \sum_{i \neq a} \frac{e_i e_a}{r_{ia}} + \sum_{i \neq a} \sum_{j>i} \frac{e_i e_j}{r_{ij}}$$

$$= \frac{1}{2M} \mathbf{K} \cdot \mathbf{K} + \sum_{i \neq a} \frac{1}{2m_i} \mathbf{p}_{ia} \cdot \mathbf{p}_{ia} + \frac{1}{2m_a} \sum_{i \neq a} \sum_{j \neq a} \mathbf{p}_{ia} \cdot \mathbf{p}_{ja} + \sum_{i \neq a} \frac{e_i e_a}{r_{ia}}$$

$$+ \sum_{i \neq a} \sum_{j>i} \frac{e_i e_j}{r_{ij}}$$

The first term of the *rhs* of this equation is constant and therefore one can use the operator

$$\mathcal{H} = \sum_{i \neq a} \frac{1}{2m_i} \mathbf{p}_{ia} \cdot \mathbf{p}_{ia} + \frac{1}{2m_a} \sum_{i \neq a} \sum_{j \neq a} \mathbf{p}_{ia} \cdot \mathbf{p}_{ja} + \sum_{i \neq a} \frac{e_i e_a}{r_{ia}} + \sum_{i \neq a} \sum_{j>i} \frac{e_i e_j}{r_{ij}}$$

The logical choice for particle *a* is the nucleus. Denoting by *m*, *e*, and *Z* the mass of the electron, the charge (in absolute value) of the electron, and the charge (in *e*-units) of the nucleus, respectively, the above equation may be rewritten (Bethe and Salpeter 1957) as

$$\mathcal{H} = \mathcal{H}_e + \mathcal{H}_{sm}$$

where

$$\mathcal{H}_e = \frac{1}{2\mu} \sum_i \mathbf{p}_{ia} \cdot \mathbf{p}_{ia} - Z \sum_i \frac{e^2}{r_{ia}} + \sum_{i<j} \frac{e^2}{r_{ij}} \tag{2.22a}$$

$$\mathcal{H}_{sm} = \frac{1}{m_a} \sum_{i<j} \mathbf{p}_{ia} \cdot \mathbf{p}_{ja} \tag{2.22b}$$

are the *electronic* and *specific mass* Hamiltonian operators, respectively, and

$$\mu = \frac{mm_a}{m + m_a}$$

is the reduced mass of the electron. The summations in Eqs. (2.22) run only over the electrons.

Using the expansion

$$\mu = \frac{m}{1 + (m/m_a)} = m\left\{ 1 - \frac{m}{m_a} + \left(\frac{m}{m_a}\right)^2 - ... \right\}$$

and taking into account that $m_a \gg m$, so that $\mu \simeq m$, Eq. (2.22a) becomes

$$\mathcal{H}_e = \frac{1}{2m} \sum_i \mathbf{p}_{ia} \cdot \mathbf{p}_{ia} - Z \sum_i \frac{e^2}{r_{ia}} + \sum_{i<j} \frac{e^2}{r_{ij}} = -\frac{\hbar^2}{2m} \sum_i \nabla_i^2 - Z \sum_i \frac{e^2}{r_{ia}} + \sum_{i<j} \frac{e^2}{r_{ij}}$$

which, in atomic units ($m = e = \hbar = 1$), is then

$$\mathcal{H}_e = \frac{1}{2} \sum_i \nabla_i^2 - Z \sum_i \frac{1}{r_{ia}} + \sum_{i<j} \frac{1}{r_{ij}} \tag{2.23}$$

The so-called *normal mass correction* is required in order to compensate for the error introduced when assuming $\mu \simeq m$. This correction is introduced, when converting the energy results to cm^{-1}, by using the conversion factor

$$R_{m_a} \simeq (1 - \frac{m}{m_a}) R_\infty$$

where R_∞ is the Rydberg constant for infinite nuclear mass.

Therefore, the results obtained with the electronic Hamiltonian operator, Eq. (2.23), will be affected by the neglect of the normal, specific, and higher-order mass corrections as well as by the neglect of the internal magnetic corrections. [For more details, numerical values, and pertinent references the reader is referred to the work of Grotch and Hegstrom (1971), Hegstrom (1973), and Fraga *et al.* (1976, 1993).]

2.4.2 Molecules

For the study of the motion of the electrons in molecules, the Hamiltonian operator given by Eq. (2.21) is customarily simplified (in atomic units) to

$$\mathcal{H}_e = \frac{1}{2} \sum_i \nabla_i^2 - \sum_i \sum_k \frac{Z_k}{r_{ik}} + \sum_{i<j} \frac{1}{r_{ij}} \tag{2.24}$$

on the basis of the Born-Oppenheimer approximation (Born and Oppenheimer 1927, Born and Huang 1955). The corresponding molecular Hamiltonian operator is then given by

$$\mathcal{H}_m = \mathcal{H}_e + \sum_{k<\ell} \frac{Z_k Z_\ell}{r_{k\ell}} \tag{2.25}$$

where the summations over k, ℓ extend to all the nuclei.

The simplification seems quite straightforward but, in fact, the implications of the use of the above operators require a detailed discussion, which is presented in Chapter 3.

2.5 The Invariant Form of the Hamiltonian Operator

The Hamiltonian operator has been presented in a general form, without any explicit mention of the system of coordinates used. One of the exceptions is Eq. (1.11), where the Laplacian operator is given explicitly in Cartesian coordinates.

In actual calculations diverse coordinate systems may be used for practical reasons. Those of interest, in particular, are the orthogonal coordinate systems (cylindrical polar, spheroidal polar, parabolic, confocal elliptic, spheroidal, parabolic cylinder, elliptic cylinder, ellipsoidal, confocal parabolic, etc.) in which the 3-dimensional Schrödinger equation can be separable (Robertson 1928, Eisenhart 1934). The results should be independent of the coordinate system and therefore it is convenient to obtain the expression of the Hamiltonian operator in terms of generalized coordinates (Schrödinger 1926). The transformation of the potential term does not offer any difficulty and therefore we will discuss only the transformation of the Laplacian.

The basic idea is already present in the transformation of the expression of the classical kinetic energy for a system of n particles, not necessarily identical. The formulation is simplified if one defines a new set of coordinates (X), through the relationships

$$X_1 = (m_1/2)^{1/2}\,x_1 \qquad\qquad X_2 = (m_1/2)^{1/2}\,y_1 \qquad\qquad X_3 = (m_1/2)^{1/2}\,z_1$$

$$X_4 = (m_2/2)^{1/2}\,x_2 \qquad\qquad X_5 = (m_2/2)^{1/2}\,y_2 \qquad\qquad X_6 = (m_2/2)^{1/2}\,z_2$$

- - - -

where m_i denotes the mass of the i-th particle and x, y, z denote the Cartesian coordinates. One can then write.

$$dX_i = \sum_k \left(\frac{\partial X_i}{\partial q_k}\right) dq_k$$

in terms of the generalized coordinates q and obtain

$$\dot{X}_i = \frac{dX_i}{dt} = \sum_k \left(\frac{\partial X_i}{\partial q_k}\right)\frac{dq_k}{dt} = \sum_k \left(\frac{\partial X_i}{\partial q_k}\right)\dot{q}_k$$

so that

$$T = \sum_i \dot{X}_i^2 = \sum_i \sum_{k,\ell} \left(\frac{\partial X_i}{\partial q_k}\right)\left(\frac{\partial X_i}{\partial q_\ell}\right)\dot{q}_k\,\dot{q}_\ell = \sum_{k,\ell} g_{k\ell}\,\dot{q}_k\dot{q}_\ell$$

with

$$g_{k\ell} = \sum_i \left(\frac{\partial X_i}{\partial q_k}\right)\left(\frac{\partial X_i}{\partial q_\ell}\right) \qquad\qquad (2.26)$$

Next we will develop the transformation of the Laplacian to a set of orthogonal curvilinear coordinates (see, e.g., Spiegel 1963) on the basis of the theory for partial derivatives. For a vector $\mathbf{r}$, with Cartesian components x, y, z, one can write

$$\mathbf{r} = x\mathbf{i} + y\mathbf{j} + z\mathbf{k} = f_x(q_1,q_2,q_3)\mathbf{i} + f_y(q_1,q_2,q_3)\mathbf{j} + f_z(q_1,q_2,q_3)\mathbf{k}$$

where the functions f define the transformation between the Cartesian coordinates and the curvilinear coordinates (q_1,q_2,q_3). One can then write

$$d\mathbf{r} = \left(\frac{\partial \mathbf{r}}{\partial q_1}\right)dq_1 + \left(\frac{\partial \mathbf{r}}{\partial q_2}\right)dq_2 + \left(\frac{\partial \mathbf{r}}{\partial q_3}\right)dq_3 = h_1 dq_1 \mathbf{u}_1 + h_2 dq_2 \mathbf{u}_2 + h_3 dq_3 \mathbf{u}_3$$

where

$$h_k = \left\|\left(\frac{\partial \mathbf{r}}{\partial q_k}\right)\right\|$$

denotes the magnitude of the vector $(\partial \mathbf{r}/\partial q_k)$ and the unit vectors in the curvilinear system are represented by $\mathbf{u}_k$.

For a given scalar function ϕ one can write

$$d\phi = \left(\frac{\partial \phi}{\partial x}\right) dx + \left(\frac{\partial \phi}{\partial y}\right) dy + \left(\frac{\partial \phi}{\partial z}\right) dz$$

which may be reexpressed as

$$d\phi = \left(\frac{\partial \phi}{\partial x}\mathbf{i} + \frac{\partial \phi}{\partial y}\mathbf{j} + \frac{\partial \phi}{\partial z}\mathbf{k}\right) \bullet (dx\,\mathbf{i} + dy\,\mathbf{j} + dz\,\mathbf{k})$$

$$= \nabla\phi \bullet d\mathbf{r} = \nabla\phi \bullet (h_1 dq_1 \mathbf{u}_1 + h_2 dq_2 \mathbf{u}_2 + h_3 dq_3 \mathbf{u}_3)$$

(2.27)

Alternatively we could write

$$d\phi = \left(\frac{\partial \phi}{\partial q_1}\right) dq_1 + \left(\frac{\partial \phi}{\partial q_2}\right) dq_2 + \left(\frac{\partial \phi}{\partial q_3}\right) dq_3$$

(2.28)

so that comparison of Eqs. (2.27) and (2.28) yields

$$\left(\frac{\partial \phi}{\partial q_1}\right) = \nabla\phi \bullet h_1 \mathbf{u}_1 \qquad \left(\frac{\partial \phi}{\partial q_2}\right) = \nabla\phi \bullet h_2 \mathbf{u}_2 \qquad \left(\frac{\partial \phi}{\partial q_3}\right) = \nabla\phi \bullet h_3 \mathbf{u}_3$$

which leads to

$$\nabla\phi = \frac{1}{h_1}\frac{\partial \phi}{\partial q_1}\mathbf{u}_1 + \frac{1}{h_2}\frac{\partial \phi}{\partial q_2}\mathbf{u}_2 + \frac{1}{h_3}\frac{\partial \phi}{\partial q_3}\mathbf{u}_3$$

Finally one obtains

$$\nabla^2\phi = \nabla \bullet (\nabla\phi) = \frac{1}{h_1}\frac{\partial}{\partial q_1}\left(\frac{1}{h_1}\frac{\partial \phi}{\partial q_1}\right) + \frac{1}{h_2}\frac{\partial}{\partial q_2}\left(\frac{1}{h_2}\frac{\partial \phi}{\partial q_2}\right) + \frac{1}{h_3}\frac{\partial}{\partial q_3}\left(\frac{1}{h_3}\frac{\partial \phi}{\partial q_3}\right)$$

which may be rewritten as

$$\nabla^2\phi = \frac{1}{h_1 h_2 h_3}\left\{ \frac{\partial}{\partial q_1}\left(\frac{h_2 h_3}{h_1}\frac{\partial \phi}{\partial q_1}\right) + \frac{\partial}{\partial q_2}\left(\frac{h_3 h_1}{h_2}\frac{\partial \phi}{\partial q_2}\right) + \frac{\partial}{\partial q_3}\left(\frac{h_1 h_2}{h_3}\frac{\partial \phi}{\partial q_3}\right) \right\}$$

(2.29)

taking into account that $\partial h_k/\partial q_\ell$ vanishes for $k \neq \ell$.

Equation (6.8) gives the Laplacian for a single particle and the corresponding kinetic energy term in the Hamiltonian operator will be $-(\hbar^2/2m)\nabla^2\phi$. It is still possible, however, to obtain a more compact expression, of interest because the procedure to obtain it and the result will help in understanding the more general approach based on tensor analysis (see below).

First we define, in a way similar to that used when transforming the expression of the classical kinetic energy, the coordinates

$$X_1 = \sqrt{m}\ x \qquad\qquad X_2 = \sqrt{m}\ y \qquad\qquad X_3 = \sqrt{m}\ z$$

We then consider a matrix **g**, with elements

$$g_{k\ell} = \sum_i \left(\frac{\partial X_i}{\partial q_k}\right)\left(\frac{\partial X_i}{\partial q_\ell}\right)$$

which are easily seen to be given by

$$g_{k\ell} = \frac{\partial(X_1 \mathbf{i} + X_2 \mathbf{j} + X_3 \mathbf{k})}{\partial q_k} \bullet \frac{\partial(X_1 \mathbf{i} + X_2 \mathbf{j} + X_3 \mathbf{k})}{\partial q_\ell} = m\left(\frac{\partial \mathbf{r}}{\partial q_k} \bullet \frac{\partial \mathbf{r}}{\partial q_\ell}\right) = mh_k^2 \delta_{k\ell}$$

(where $\delta_{k\ell}$ is the Kronecker delta). This matrix could have been obtained

$$\mathbf{g} = (\mathbf{g}^{1/2})^\dagger \mathbf{g}^{1/2}$$

from a matrix $\mathbf{g}^{1/2}$, with elements

$$g_{ij}^{1/2} = \frac{\partial X_i}{\partial q_j}$$

(taking into account the above result for the elements $g_{k\ell}$). [See the Appendix for the notation used for matrices.] The corresponding determinants are

$$g = \begin{vmatrix} mh_1^2 & 0 & 0 \\ 0 & mh_2^2 & 0 \\ 0 & 0 & mh_3^2 \end{vmatrix}$$

$$g^{1/2} = \begin{vmatrix} \dfrac{\partial X_1}{\partial q_1} & \dfrac{\partial X_1}{\partial q_2} & \dfrac{\partial X_1}{\partial q_3} \\[2mm] \dfrac{\partial X_2}{\partial q_1} & \dfrac{\partial X_2}{\partial q_2} & \dfrac{\partial X_2}{\partial q_3} \\[2mm] \dfrac{\partial X_3}{\partial q_1} & \dfrac{\partial X_3}{\partial q_2} & \dfrac{\partial X_3}{\partial q_3} \end{vmatrix} = \frac{\partial(X_1, X_2, X_3)}{\partial(q_1, q_2, q_3)} \qquad\qquad (2.30)$$

with

$$g^{1/2} = \sqrt{g}$$

(taking into account that the determinant of a matrix obtained as the product of two other matrices is equal to the product of their determinants). The determinant $g^{1/2}$ is the Jacobian, relating the two sets of coordinates, and the *rhs* of Eq. (2.30) represents the corresponding short notation for it.

Taking into account these developments one can see that the kinetic energy term will be given by

$$-\frac{\hbar^2}{2} \frac{1}{\sqrt{g}} \left\{ \frac{\partial}{\partial q_1} \left(\frac{\sqrt{g}}{g_{11}} \frac{\partial \Psi}{\partial q_1} \right) + \frac{\partial}{\partial q_2} \left(\frac{\sqrt{g}}{g_{22}} \frac{\partial \Psi}{\partial q_2} \right) + \frac{\partial}{\partial q_3} \left(\frac{\sqrt{g}}{g_{33}} \frac{\partial \Psi}{\partial q_3} \right) \right\}$$

which can be rewritten as

$$-\frac{\hbar^2}{2} \frac{1}{\sqrt{g}} \left\{ \frac{\partial}{\partial q_1} \left(\sqrt{g}\, g^{11} \frac{\partial \Psi}{\partial q_1} \right) + \frac{\partial}{\partial q_2} \left(\sqrt{g}\, g^{22} \frac{\partial \Psi}{\partial q_2} \right) + \frac{\partial}{\partial q_3} \left(\sqrt{g}\, g^{33} \frac{\partial \Psi}{\partial q_3} \right) \right\}$$

where $g^{k\ell}$ is the $k\ell$-element of the matrix $\mathbf{g}^{-1}$ and is given by

$$g^{k\ell} = \frac{G_{\ell k}}{g}$$

where $G_{k\ell}$ is the cofactor of $g_{k\ell}$ in the determinant g.

The formulation could now be extended to a many-particle system but the corresponding result

$$-\frac{\hbar^2}{2} \frac{1}{\sqrt{g}} \sum_{k,\ell} \frac{\partial}{\partial q_k} \left(\sqrt{g}\, g^{k\ell} \frac{\partial \Psi}{\partial q_\ell} \right)$$

may be obtained in a more elegant way by the techniques of tensor analysis (see, e.g., McConnell 1957 and Rapp 1971), in which case the quantities $g_{k\ell}$ are considered to be the elements of the so-called metric tensor (Mott and Sneddon 1963) of the system of coordinates. Tensor analysis constitutes a very powerful technique, whereby very compact expressions are obtained, at the cost of a typographically difficult notation, with variations among the different authors, and therefore care must be exercised when comparing expressions.

As integration will be required for the calculation of expectation values, it is also necessary to obtain the expression of the volume element in the system of coordinates being considered. One will write

$d\tau = (h_1 dq_1 \mathbf{u}_1) \cdot [(h_2 dq_2 \mathbf{u}_2) \times (h_3 dq_3 \mathbf{u}_3)] = h_1 h_2 h_3 dq_1 dq_2 dq_3$

which, taking into account the preceding developments, may be rewritten as

$$d\tau = \frac{\partial(x,y,z)}{\partial(q_1,q_2,q_3)} \, dq_1 \, dq_2 \, dq_3$$

Pauling and Wilson (1935) have collected the expressions for ∇^2 and $d\tau$ for various coordinate systems and Kemble (1958) has discussed the probability density in q-space and applied the formulation to the case of spherical coordinates.

3 The Schrödinger Equation in Position Space

We will now focus our attention on the Schrödinger equation *per se* (Schrödinger 1926), with a perspective of the interdependent characteristics of the Hamiltonian operator and its eigenfunctions, an examination of some local properties, and an analysis of global properties.

There is a wealth of published material on the Schrödinger equation and, in addition to those papers explicitly mentioned within the text, the reader is referred to the many textbooks available, as listed in the bibliography section. In particular it should be mentioned that our interests preclude a discussion of the methods, techniques, and calculations which, based mainly on the perturbation formalism and the variational principle, constitute the body of the so-called computational chemistry.

For simplicity in the formulation we will use the short notation

$$\mathcal{H}\Psi_I = E_I\Psi_I \tag{3.1}$$

for the Schrödinger equation, independently of the Hamiltonian operator involved, except where the corresponding specific information is required. In particular, the label for the eigenstate and the variable dependence of Ψ and E will be often omitted.

3.1 The Hamiltonian Operator and Its Eigenfunctions

We will be considering the general operator

$$\mathcal{H} = \sum_i \left\{ \frac{1}{2m_i} \mathbf{p}_i \cdot \mathbf{p}_i + \sum_{j>i} \frac{e_i e_j}{r_{ij}} \right\} \tag{3.2}$$

for a many-electron, many-nuclei system, with the understanding that the general discussion below applies as well (unless otherwise indicated) to the case of atoms.

When the distinction between electrons and nuclei is introduced, the above equation may be written (with the notation already used in Chapter 2) as

$$\mathcal{H} = \sum_i \left\{ \frac{1}{2m} \mathbf{p}_i \cdot \mathbf{p}_i + \sum_{j>i} \frac{e^2}{r_{ij}} \right\} - \sum_i \sum_k \frac{Z_k e^2}{r_{ik}} + \sum_k \left\{ \frac{1}{2m_k} \mathbf{p}_k \cdot \mathbf{p}_k + \sum_{k>\ell} \frac{Z_k Z_\ell e^2}{r_{k\ell}} \right\}$$

with the corresponding Schrödinger equation given, in short, as

$$\mathcal{H}\Psi_I(\mathbf{r}_i,\mathbf{r}_k) = E_I \Psi_I(\mathbf{r}_i,\mathbf{r}_k) \tag{3.3}$$

where E_I and $\Psi_I(\mathbf{r}_i,\mathbf{r}_k)$ denote the eigenvalue and eigenfunction, respectively, corresponding to the I-th eigenstate. For simplicity, the eigenstate label and the dependence on the positions of the electrons ($\mathbf{r}_i$) and of the nuclei ($\mathbf{r}_k$) will be omitted hereafter except where that information is required.

The difficulties in the solution of Eq. (3.3) have led to the *clamped-nuclei* formalism (Sutcliffe 1992, 1993, 1994, 1995). Within the context of this formalism the nuclei are assumed to be at fixed (clamped) positions, defined by the position vectors $\mathbf{a}_k$, which are parameters identifying the molecular (nuclear) geometry. The complete eigenfunction is then given by

$$\Psi(\mathbf{r}_i,\mathbf{a}_k) = \sum_p \phi_p(\mathbf{a}_k)\psi_p(\mathbf{r}_i,\mathbf{a}_k) \tag{3.4}$$

where the summation over p extends to all the possible geometries, i.e., to all the possible values of the parameters $\mathbf{a}_k$. The functions $\psi(\mathbf{r}_i,\mathbf{a}_k)$ are the solutions of the eigenequation

$$\mathcal{H}_e\psi(\mathbf{r}_i,\mathbf{a}_k) = E_e(\mathbf{a}_k)\psi(\mathbf{r}_i,\mathbf{a}_k) \tag{3.5}$$

where

$$\mathcal{H}_e = \sum_i \left\{ \frac{1}{2m} \mathbf{p}_i \cdot \mathbf{p}_i - \sum_k \frac{Z_k e^2}{r_{ik}} + \sum_{j>i} \frac{e^2}{r_{ij}} \right\} \tag{3.6}$$

is the clamped-nuclei Hamiltonian operator, normally used in actual calculations, and r_{ik} is now defined as $r_{ik} = \mathbf{r}_i - \mathbf{a}_k$.

The molecular energy corresponding to a given molecular geometry is then evaluated as

$$E_m(\mathbf{a}_k) = E_e(\mathbf{a}_k) + \sum_{k<\ell} \frac{Z_k Z_\ell e^2}{r_{k\ell}} \tag{3.7}$$

with $r_{k\ell} = \mathbf{a}_k - \mathbf{a}_\ell$.

The implications of the clamped-nuclei formalism are analyzed in Section 3.3.1.

3.1.1 The Hamiltonian Operator

The basic considerations, interconnecting the Hamiltonian operator and its eigenfunctions, are the requirement that every physical observable should be represented by a self-adjoint operator in a Hilbert space of quadratically integrable functions defined in the configuration space (von Neumann 1932) and the Hermitian character of the (linear) Hamiltonian operator for a given manifold of functions with reference to a given domain of the variables.

The energy is a physical quantity and therefore it must be associated with a self-adjoint operator. If the Hamiltonian operator is adjoint to itself (that is, self-adjoint) with respect to a manifold of functions and a domain of the variables, it will be said to be Hermitian with respect to the manifold and the domain.

It can be proved (Kemble 1958) that the Hamiltonian operator is a Hermitian operator and therefore it follows from the definition that it is self-adjoint. Being self-adjoint means, in particular, that the transform $\mathcal{H}\Psi$ is quadratically integrable.

Using the customary bra-ket notation for expectation values, the Hermitian character of the Hamiltonian operator may be expressed by

$$\langle \Psi_I | \mathcal{H} \Psi_J \rangle = \langle \mathcal{H} \Psi_I | \Psi_J \rangle \qquad (3.8)$$

which will hold for every pair of functions Ψ_I and Ψ_J in the manifold when the integration is extended over the complete domain.

3.1.2. The Eigenfunctions

The Hamiltonian operator is a formal differential operator and its eigenfunctions will not be determined unless some boundary or continuity conditions at the singular points of the potential (see below) are imposed (Kato 1957).

The manifold of functions, for which the Hermitian character of the Hamiltonian was proved, has the following characteristics (Kemble 1958):

(a) They are continuous and single-valued throughout the configuration space and analytic in all the variables at every point where the potential energy is analytic.

(b) They have first-order partial derivatives (except at the Coulomb-type singular points of the potential) which are bounded (Kato 1957). They and their first-order partial derivatives must be absolutely and quadratically integrable over the whole configuration space.

(c) They are eigenfunctions of $\mathcal{H}^n$ (that is, the n-fold application of $\mathcal{H}$, n being positive).

These eigenfunctions are the so-called *genuine* eigenfunctions, corresponding to discrete (point) eigenvalues. The class of *generalized* eigenfunctions, defined by Kato (1957), includes in addition wave packets.

There may exist also *continuous* eigenfunctions [or *eigendifferentials* (Kemble 1958)], belonging to the continuous region of the energy spectrum. It can be postulated that these continuous eigenfunctions belong to the same class of eigenfunctions with respect to which the Hamiltonian operator is Hermitian.

The genuine eigenfunctions of one-electron systems (Hydrogen-like atoms and Hydrogen-ion-like molecules) are known and the existence of generalized eigenfunctions has been proved for many-particle systems (Kato 1951).

Some basic properties of the eigenvalues and eigenfunctions are as follows:

(a) Real character of the eigenvalues

Multiplication on the left of Eq. (3.1) with Ψ_I^* and integration over the whole space yields

$$\langle \Psi_I | \mathcal{H} | \Psi_I \rangle = E \langle \Psi_I | \Psi_I \rangle$$

which, upon complex conjugation, transforms into

$$\langle \Psi_I | \mathcal{H} | \Psi_I \rangle^* = E^* \langle \Psi_I | \Psi_I \rangle^* = E^* \langle \Psi_I | \Psi_I \rangle$$

Taking into account the Hermitian character of $\mathcal{H}$, one can write

$$\langle \Psi_I | \mathcal{H} | \Psi_I \rangle^* = \langle \mathcal{H} \Psi_I | \Psi_I \rangle = \langle \Psi_I | \mathcal{H} \Psi_I \rangle$$

from which it follows that $E = E^*$.

(b) Orthonormality of the eigenfunctions

For two functions Ψ_I and Ψ_J, belonging to different eigenvalues E_I and E_J, respectively, one can write, proceeding as above,

$$\langle \Psi_J | \mathcal{H} | \Psi_I \rangle = E_I \langle \Psi_J | \Psi_I \rangle \tag{3.9a}$$

$$\langle \Psi_I | \mathcal{H} | \Psi_J \rangle = E_J \langle \Psi_I | \Psi_J \rangle \tag{3.9b}$$

Taking the complex conjugate of Eq. (3.9a) we obtain

$$\langle \Psi_J | \mathcal{H} | \Psi_I \rangle^* = E_I \langle \Psi_J | \Psi_I \rangle$$

while the Hermitian character of $\mathcal{H}$ allows us to write, from Eq. (3.9b)

$$\langle\Psi_I|\mathcal{H}|\Psi_J\rangle = \langle\mathcal{H}\Psi_I|\Psi_J\rangle = \langle\Psi_J|\mathcal{H}|\Psi_I\rangle^* = E_J\langle\Psi_I|\Psi_J\rangle$$

so that

$$(E_I - E_J)\langle\Psi_I|\Psi_J\rangle = 0$$

With $E_I \neq E_J$, it must be $\langle\Psi_I|\Psi_J\rangle = 0$.

In the case of degenerate eigenfunctions, that is, eigenfunctions associated with the same eigenvalue, they are either orthogonal to each other or linear combinations of them may be formed which will be orthogonal.

The eigenfunctions of the Hamiltonian operator are not normalized necessarily but the condition

$$\langle\Psi_I|\Psi_I\rangle = 1$$

may be imposed, if desired, without affecting the energy expectation value. Hereafter it will be assumed that the eigenfunctions are normalized.

Equivalent orthonormality conditions may be imposed for the continuous eigenfunctions (Kemble 1958).

(c) Completeness of the set of eigenfunctions

One can postulate the completeness of the set of eigenfunctions of the Hamiltonian operator, including the continuous eigenfunctions if they exist. Any arbitrary function may then be expanded (Kemble 1958) as

$$\psi_k = \sum_I \Psi_I c_{Ik} + \sum_K \int \Psi_K(E)c_{Kk}(E)dE$$

where the summations over I and K extend over the genuine and continuous eigenfunctions, respectively, and c_{Ik}, $c_{Kk}(E)$ are appropriate coefficients.

If the genuine eigenfunctions form a complete set by themselves, the corresponding expansion becomes simply

$$\psi_k = \sum_I \Psi_I c_{Ik} \tag{3.10}$$

If the functions ψ form a orthonormal set and taking into account the orthonormality of the functions Ψ one can write

$$\langle\psi_k|\psi_k\rangle = \sum_I \sum_J c_{Jk}^* \langle\Psi_J|\Psi_I\rangle c_{Ik} = \sum_I c_{Ik}^* c_{Ik} = 1$$

$$\langle\psi_\ell|\psi_k\rangle = \sum_I \sum_J c_{J\ell}^* \langle\Psi_J|\Psi_I\rangle c_{Ik} = \sum_I c_{I\ell}^* c_{Ik} = 0$$

Equation (3.10) may be rewritten as

$$\psi_k = \Psi c_k \tag{3.11}$$

where Ψ is a row vector, with elements Ψ_1, Ψ_2, ..., and c_k is a column vector, with elements c_{1k}, c_{2k}, ... For the complete set of functions ψ, the set of Eqs. (3.11) may be combined into

$$\psi = \Psi C$$

where ψ is a row vector, with elements ψ_1, ψ_2, ..., and C is a matrix formed by the column vectors c_1, c_2, ...

If the two sets consist of the same number of functions, C is then a square matrix. In such a case one can write

$$\psi C^{-1} = \Psi C C^{-1} = \Psi$$

for the reverse expansion of the functions Ψ in terms of the functions ψ. The matrix C is therefore either an orthogonal matrix ($C^{-1} = C^\dagger$), if the expansion coefficients are real, or a unitary matrix ($C^{-1} = \overline{C}$), if the expansion coefficients are complex. [See the Appendix for the notation used for matrices.]

(d) Eigenfunctions of commuting operators.

Let us consider the set Ψ of eigenfunctions of the Hamiltonian operator and a set of functions ψ, which are eigenfunctions

$$\mathcal{P}\psi_k = p_k\psi_k$$

of the operator $\mathcal{P}$, which commutes with $\mathcal{H}$. Expanding ψ_k in terms of the set Ψ, Eq. (3.10), and operating with $\mathcal{P}$ one obtains

$$\mathcal{P}\psi_k = \sum_I \mathcal{P}\Psi_I c_{Ik} = p_k \sum_I \Psi_I c_{Ik}$$

which can be rewritten as

$$\sum_I (\mathcal{P} - p_k)\Psi_I c_{Ik} = 0 \tag{3.12}$$

Every term of this expansion transforms, under $\mathcal{H}$, as

$$\mathcal{H}(\mathcal{P} - p_k)\Psi_I c_{Ik} = (\mathcal{P} - p_k)\mathcal{H}\Psi_I c_{Ik} = E_I(\mathcal{P} - p_k)\Psi_I c_{Ik}$$

which shows that $(\mathcal{P} - p_k)\Psi_I c_{Ik}$ is an eigenfunction of $\mathcal{H}$ with an eigenvalue E_I; therefore it must be proportional, $\kappa_{kI}\Psi_I$, to Ψ_I. Equation (3.12) may then be rewritten as

$\sum_I \kappa_{kI} \Psi_I c_{Ik} = 0$

so that multiplication on the left with Ψ_J^* and integration yields

$\sum_I \kappa_{kI} <\Psi_J|\Psi_I> c_{Ik} = \kappa_{kJ} c_{Jk} = 0$

That is, $\kappa_{kJ} = 0$ for every J. Therefore one has

$(\mathcal{P} - p_k)\Psi_I c_{Ik} = 0$

or, equivalently,

$\mathcal{P}\Psi_I = p_k \Psi_I$

which shows that the eigenfunctions of $\mathcal{H}$ are simultaneously eigenfunctions of an operator $\mathcal{P}$, which commutes with $\mathcal{H}$.

This result is of particular interest in connection with the angular momentum operators but here we will use it in order to discuss the symmetry properties of the eigenfunctions Ψ.

(e) Symmetry of the eigenfunctions.

The two types of operators that we want to consider here are the permutation operator and the symmetry operators from group theory.

The permutation operator $\mathcal{P}_{ij}$, when acting on $\mathcal{H}$, will exchange the electrons i and j. Because the summations over i, j in $\mathcal{H}$ extend to all the electrons and the interelectronic terms involve only the scalar quantity r_{ij}, the Hamiltonian operator will remain unchanged. The action of $\mathcal{P}_{ij}^2$ on $\mathcal{H}\Psi_I$ may be written as

$\mathcal{P}_{ij}^2 \mathcal{H}\Psi_I = (\mathcal{P}_{ij}^2 \mathcal{H})\Psi_I = \mathcal{H}\Psi_I = E_I \Psi_I$

or

$\mathcal{P}_{ij}^2 (\mathcal{H}\Psi_I) = \mathcal{P}_{ij}^2 (E_I \Psi_I) = E_I \mathcal{P}_{ij}^2 \Psi_I$

so that it must be

$\mathcal{P}_{ij}^2 \Psi_I = \Psi_I$ \hfill (3.13)

That is: the eigenfunctions Ψ_I are eigenfunctions of the operator $\mathcal{P}_{ij}^2$, with an eigenvalue of unity.

If we denote by ψ_k the eigenfunctions of $\mathcal{P}_{ij}$, one will have

$$\mathcal{P}_{ij}\psi_k = p_k\psi_k$$

and

$$\mathcal{P}_{ij}^2\psi_k = \mathcal{P}_{ij}(p_k\psi_k) = p_k\mathcal{P}_{ij}\psi_k = p_k^2\psi_k \tag{3.14}$$

so that comparison of Eqs. (3.13) and (3.14) shows that $p_k = \pm 1$. That is: the eigenvalues of the permutation operator are ± 1 and therefore the eigenfunctions Ψ may be either *symmetric* or *antisymmetric* with respect to the exchange of any two electrons.

The additional postulate of quantum mechanics, embodied in the Pauli principle, establishes that the eigenfunctions of a many-electron system must be antisymmetric with respect to the exchange of any two electrons.

A symmetry group consists of those operations, with respect to given elements of symmetry, which transform a geometric figure into itself, leaving all the distances between pairs of points in the figure unchanged. The elements of symmetry for molecules are axes of symmetry, planes of symmetry, centre of symmetry, and so-called rotation-reflection axes of symmetry. Therefore, the operations to be considered are rotations, reflections, inversion, and rotation-reflections, as well as the identity operation (in which no transformation is performed). Associated with each symmetry operation there exists a symmetry operator.

Within the context of the clamped nuclei approximation, the symmetry operations leave unchanged the Hamiltonian operator and the symmetry operators commute with the latter (Schonland 1965). The eigenfunctions Ψ constitute bases for the irreducible representations of the symmetry group corresponding to the molecular symmetry (defined by the geometric figure formed by the nuclei); in particular, eigenfunctions corresponding to degenerate eigenvalues will constitute bases for irreducible representations of order $n > 1$.

The practical application of group theory in computational chemistry lies outside of the scope of this work and therefore the reader is referred, e.g., to the works of Cotton (1970) and Schonland (1965). [The work of Wherrett (1986) contains, in addition, a graded listing of the most important textbooks on the subject.]

Section 4.4.1 presents some considerations of the symmetry in the p-representation.

3.2 Local Properties

Although our main interest is the analysis of local energies (including the consideration of the regions of null kinetic and null potential energy), some brief comments must be made regarding the existence of nodes and nodal planes.

3.2.1 Nodes and Nodal Planes

Radial nodes/nodal planes are those points/planes where an eigenfunction vanishes. For example, within the context of the Schrödinger theory considered here, the s-type eigenfunctions (other than the ground state eigenfunction) of Hydrogen-like atoms have radial nodes. Their existence would pose a paradox, which should be explained on the basis of the uncertainty principle and of a redefinition of matter (Kemble 1958).

It must be remembered, first, that the Schrödinger theory represents only an approximation to reality and that a proper analysis of the problem should be carried out within the context of relativistic theory. Thus, for the state $2S_{1/2}$ (i.e., with principal, orbital angular momentum, and total electron angular momentum quantum numbers $n = 2$, $\ell = 0$, $j = 1/2$) of Hydrogen-like atoms the radial functions are of the form

$$g(\rho) = k_g \left\{ N - \frac{N+1}{2\gamma+1} \rho \right\} \rho^{\gamma-1} e^{-1/2\rho} \qquad (3.15a)$$

$$f(\rho) = k_f \frac{(2\gamma+1)(N+2) - (N+1)\rho}{(2\gamma+1)N - (N+1)\rho} g(\rho) \qquad (3.15b)$$

with

$$N = [2(1+\gamma)]^{1/2} \qquad \gamma = [1-(\alpha Z)^2]^{1/2} \qquad \rho = \frac{2Z}{N} r$$

and where k_g, k_f are constants and α is the fine-structure constant. [For more details see the work of Bethe and Salpeter (1957) and the original work of Darwin (1928), Gordon (1928), White (1931), Burhop and Massey (1935), and Davis (1939)].

Substitution of the expression for $g(\rho)$ into Eq. (3.15b) yields

$$f(\rho) = k_f k_g \frac{(2\gamma+1)(N+2) - (N+1)\rho}{2\gamma+1} \rho^{\gamma-1} e^{-1/2\rho}$$

The functions $g(\rho)$ and $f(\rho)$ vanish, respectively, at the points

$$\rho_g = \frac{N(2\gamma+1)}{N+1} \qquad \rho_f = \frac{(N+2)(2\gamma+1)}{N+1} = \rho_g + \frac{2(2\gamma+1)}{N+1}$$

which are different. That is, the four spinors of the eigenfunction will not vanish simultaneously at the same point.

Similarly, within the context of the Schrödinger theory the p-type eigenfunctions present nodal planes because of the spherical harmonics (with the same ℓ value) in the angular part of the function. Within the context of relativistic theory, the spinors for a given ℓ will depend on $Y_{\ell,m\pm 1/2}$ and $Y_{\ell+1,m\pm 1/2}$ (for $j = \ell + 1/2$) and $Y_{\ell,m\pm 1/2}$ and $Y_{\ell-1,m\pm 1/2}$ (for $j = \ell - 1/2$). The symmetry characteristics of the spherical harmonics involved (with ℓ and $\ell+1$ or ℓ and $\ell-1$) will determine again that the four spinors will not share a common nodal plane.

In the case of molecules we have already mentioned in the preceding section that the eigenfunctions of the clamped-nuclei Hamiltonian operator are bases for the irreducible representations of the symmetry group for the molecular geometry under consideration. For nuclear configurations with a high degree of symmetry, some eigenfunctions will present nodal planes. However, the use of the clamped-nuclei Hamiltonian operator represents only an approximation and the complete function must be obtained through consideration of all possible geometries, as implied by Eq. (3.3). Therefore, the complete eigenfunction $\Psi(r_i,r_k)$ will not present nodal planes.

There is no need for further discussion of these points but the interested reader may wish to consult, for example, the work of Moss (1973) and references therein for further details.

3.2.2 Local Energies

Equation (3.1) may be rewritten, for any eigenstate, as

$$E = \frac{\mathcal{H}\Psi}{\Psi} \tag{3.16}$$

which indicates that the ratio of the transform $\mathcal{H}\Psi$ over Ψ, at any point in the electron configuration space, must be equal to the value of the energy of the eigenstate under consideration (see Section 5.2). In particular, this must be true at those points where the Hamiltonian operator presents singularities; that is, it must be

$$\left(\frac{\mathcal{H}\Psi}{\Psi}\right)_{r=0} = E \tag{3.17}$$

where r denotes the interparticle coalescence coordinate.

The problem may be reformulated as follows. Differentiating Eq. (3.16) with respect to r and reordering yields, for $r = 0$,

$$\left(\frac{1}{\mathcal{H}\Psi}\frac{\partial \mathcal{H}\Psi}{\partial r}\right)_{r=0} = \left(\frac{1}{\Psi}\frac{\partial \Psi}{\partial r}\right)_{r=0} \tag{3.18}$$

which allows us to study the behaviour of the *lhs* of this equation by looking at the behaviour of the *rhs*. This problem has been studied by Kato (1957), whose work is briefly summarized below. [From T. Kato, *On the eigenfunctions of many-particle systems in Quantum Mechanics*, in *Communications on Pure and Applied Mathematics.* vol. 10, pp. 151-177 (1957). Copyright© 1957 John Wiley & Sons, Inc., New York, U.S.A. Reprinted by permission of John Wiley & Sons, Inc.]

Let us consider a potential of the general form

$$V(r_1,...,r_n) = V_0(r_1,...,r_n) + \sum_i V_{0i}(r_i) + \sum_{i<j} V_{ij}(r_{ij}) \qquad (3.19)$$

where the summations over i,j extend to the n particles (including both electrons and nuclei) of the system. V_0 is a real valued, measurable function bounded in the whole $3n$-configuration space and V_{ij}, $0 \le i < j \le n$, is a real valued, measurable function, defined in the 3-dimensional space r, which vanishes identically outside some sphere and satisfies the condition

$$\int \left| V_{ij}(r) \right|^\sigma dr < \infty \qquad (3.20)$$

where σ is a fixed constant (≥ 2). These conditions are satisfied by a Coulomb potential.

The Coulomb potential will be denoted as generalized if it can be expressed in the form of Eq. (3.19), with V_0 satisfying the same condition as above and if, for each $0 \le i < j \le n$, $V_{ij}(r)$ may be expressed as

$$V_{ij}(r) = k_{ij} \frac{f(r)}{r} + V'_{ij}(r) \qquad (3.21)$$

where k_{ij} is a constant, $f(r)$ is a function equal to 1 for $r \le \rho$, $\rho > 0$, and equal to 0 otherwise, and the $V'_{ij}(r)$ satisfy the condition given by Eq. (3.20) with $\sigma > 3$.

Given the potential V, Eq. (3.19), all the eigenfunctions are bounded and are uniformly Hölder continuous functions, with $\theta < 2 - 3/\sigma$; if $\sigma > 3$, all the first-order partial derivatives of the eigenfunctions exist and are uniformly Hölder continuous functions, with $\theta < 1 - 3/\sigma$. [A function ϕ is said to be uniformly Hölder continuous, with exponent θ, if the constants C and θ, $0 < \theta < 1$, exist such that $|\phi(p) - \phi(q)| \le C(\overline{pq})^\theta$, where p and q are any two points in the configuration space and $\overline{pq}$ denotes the distance between them.] Given the generalized Coulomb potential, all the eigenfunctions are continuous and have derivatives of first order, except at the Coulomb singular points, and those derivatives are bounded.

These results may now be supplemented as follows. A linear transformation of the form

$$r'_i = \sum_k a_{ik} r_k \qquad (3.22)$$

where the subscript i labels the i-th particle and the summation extends to the n particles, transforms each of the three sets of coordinates (x_i), (y_i), and (z_i) into themselves and in the same way. In the new coordinate system, $(x_1, x_2, ..., x_m)$, $m = 3n$, the Laplacian is expressed in terms of the derivatives $\partial^2/\partial x_i^2$ and the potential becomes

$$\mathcal{V}(x) = \mathcal{V}_0(x) + \sum_i \mathcal{V}_i(x) \tag{3.23}$$

with i running from 1 to $\ell = \frac{1}{2}n\,(n + 1)$. An orthogonal transformation may be found that will yield a new coordinate system, $(X_1, X_2, ..., X_m)$, such that the Laplacian remains unchanged and the terms $\mathcal{V}_i(x)$ are transformed into $\mathcal{V}_i(X_1, X_2, X_3)$, the so-called canonical form. This potential has the same properties as $\mathcal{V}_{ij}(r)$ and may be expressed as

$$\mathcal{V}_i(X_1, X_2, X_3) = k_i\,\frac{f(r)}{r} + \mathcal{V}_i'(X_1, X_2, X_3) \tag{3.24}$$

where k_i and $f(r)$ have the same meaning as above, $r = (X_1^2 + X_2^2 + X_3^2)^{1/2}$, and $\mathcal{V}_i(X_1, X_2, X_3)$ satisfies the condition

$$\int \left| \mathcal{V}_i'(X_1, X_2, X_3) \right|^\sigma dX_1, dX_2, dX_3 < \infty \tag{3.25}$$

(with $\sigma > 3$). The designation of internal and external coordinates, with respect to $\mathcal{V}_i$, is given to (X_1, X_2, X_3) and $(X_4, ..., X_m)$, respectively.

Then, given the potential $\mathcal{V}$, Eq. (3.19), any eigenfunction Ψ is the sum of ℓ functions ψ such that, for each i, ψ_i is a uniformly Hölder continuous function, with $\sigma > 2$, $\theta < 2-3/\sigma$, and all the external derivatives of ψ_i, with respect to $\mathcal{V}_i$, exist and are uniformly Hölder continuous functions, with $\theta < 3-6/\sigma$. For the generalized Coulomb potential, any eigenfunction Ψ is the sum of $\ell+1$ functions $\psi_0, \psi_1, ..., \psi_\ell$ such that all the first-order partial derivatives of ψ_0 exist and are uniformly Hölder continuous functions, with $\theta < 1-3\sigma$, and that, for the canonical coordinate system for $\mathcal{V}_i$, ψ_i is a function of only $(m\text{-}2)$ variables $r = \left(X_1^2 + X_2^2 + X_3^2 \right)^{1/2}, X_4, ...,$ X_m, and that its first-order partial derivatives exist and are uniformly Hölder continuous functions, with $\theta < 1$.

Taking into account that $r = \left(X_1^2 + X_2^2 + X_3^2 \right)^{1/2}$, one can see that the internal derivatives

$$\frac{\partial \psi_i}{\partial X_k} = \frac{\partial \psi_i}{\partial r}\frac{\partial r}{\partial X_k} = \frac{X_k}{r}\frac{\partial \psi_i}{\partial r} \qquad k = 1, 2, 3 \tag{3.26}$$

are continuous for $r > 0$ but discontinuous for $r = 0$. This discontinuity is transferred over to the internal derivatives of Ψ, with respect to $\mathcal{V}_i$, because all the other functions ψ_j, $j \neq i$, are continuously differentiable at the point $(0, 0, 0, X_4, ..., X_m)$, unless this point is a Coulomb singular point for another term of the potential.

The internal derivatives of Ψ, for $r \to 0$, behave as

$$\frac{\partial \Psi}{\partial X_k} = a_{ik}(X_4, ..., X_m) + \frac{X_k}{r}\left(\frac{\partial \psi_i}{\partial r}\right)_{r=0} + O(r^\theta) \qquad (3.27)$$

from which it can be seen that

$$\left(\frac{\partial \psi_i}{\partial r}\right)_{r=0} = \left(\frac{\partial \Psi_{av}}{\partial r}\right)_{r=0}$$

where Ψ_{av} represents the average value of Ψ over a sphere r = constant, for fixed values of $X_4, ..., X_m$. This quantity is proportional

$$\left(\frac{\partial \Psi_{av}}{\partial r}\right)_{r=0} = \frac{1}{2} k_i \, \Psi(0, 0, 0, X_4, ..., X_m) \qquad (3.28)$$

to $\Psi(0, 0, 0, X_4, ..., X_m)$; k_i depends on the form of the Hamiltonian operator as well as on r.

3.2.3 Coalescence and Cusp Conditions

Equation (3.28), which is customarily written as

$$\left(\frac{1}{\Psi}\frac{\partial \Psi_{av}}{\partial r}\right)_{r=0} = \zeta \qquad (3.29)$$

represents a *cusp* condition, which must be satisfied by the eigenfunctions of the Schrödinger equation. Cusp conditions constitute particular cases of *coalescence* conditions, when the eigenfunction does not vanish at the coalescence of the two particles.

For a many-electron atom (with nuclear charge Z), using the non-relativistic Hamiltonian operator within the context of the (heavy) fixed-nucleus approximation, ζ takes the values $-Z$ and $1/2$, respectively, for the electron-nucleus and electron-electron coalescence (Roothaan and Weiss 1960).

A corollary, derived by Steiner (1963),

$$\left(\frac{1}{\rho(r)}\frac{\partial \rho(r)}{\partial r}\right)_{r=0} = -2Z$$

relates the charge density $\rho(r)$ and its derivative at the nucleus. The implications of this condition in practical calculations, within the framework of the expansion approximation, have been discussed by Fraga and Malli (1968).

Because of its importance, a considerable effort has been dedicated to the topic of cusp conditions. Some of the points investigated are the electron-electron cusp condition in H_2 (Kolos and Roothaan 1960), the electron-nucleus cusp conditions in polyatomic molecules (Bingel 1963), the electron-nucleus cusp condition for a general ℓ-wave spherical harmonic component in an atomic function (Roothaan and Kelly 1963), the effect of removing the fixed-nucleus approximation and of not invoking the spherical averaging (Pack and Byers Brown 1966), the electron-electron cusp condition in a uniform electron gas (Kimball 1973, 1975), the general ℓ-wave conditions for 2-electron density matrices (Davidson 1976), the representation of cusps in a hyperspherical basis set (Avery and Antonsen 1996), and the relationship of the third derivative of the spherically-averaged function with the lower derivatives (Rassolov and Chipman 1996). Additional work has been carried out on the relationship with the united-atom formalism for polyatomic molecules (Bingel 1963), the correlation between the electron-nucleus cusp condition and the accuracy of the spin density at the nucleus (Chang *et al.* 1970, Chapman and Chong 1970), the relation between the cusp behaviour and the structure of a uniform electron gas (Kimball 1973, Rajogopal *et al.* 1978), the convergence of configuration interaction functions (Kutzelnigg and Klopper 1991), and a spatial generalization for 2-electron atoms with correlation (Arias de Saavedra *et al.* 1996), as well as other diverse practical applications (Roothaan and Kelly 1963, Chong 1967, and Poling *et al,* 1971).

Special mention must be made of the detailed analysis of the problem of cusp conditions carried out by King (1996), with a renewed examination of the different practical relevance of 2- and 3-body cusp conditions in conjunction with energy calculations (Roothaan and Weiss 1960). On one hand, the electron-nucleus cusp condition is important in an energy calculation but it is easily satisfied (see, e.g., Fraga and Malli 1968). On the other hand, the 3-body cusp conditions (for the coalescence of either three electrons or two electrons and a nucleus) are difficult to satisfy but they are not important for energy calculations.

3.2.4 Null Kinetic- and Potential-Energy Regions

While the value of the total energy must be a constant at each point in the electron configuration space, this is not the case for either the kinetic or the potential energy. Within the framework of the Schrödinger theory there are, however, some characteristics of interest regarding these two energies.

The clamped-nuclei eigenequation may be written in abbreviated form

$$(\mathcal{T} + \mathcal{V})\Psi = E\Psi$$

where $\mathcal{T}$ and $\mathcal{V}$ denote the kinetic and the potential energy operators, respectively. This equation may be rewritten as

$$\mathcal{V}\Psi = (E - \mathcal{T})\Psi \tag{3.30a}$$

or

$$\mathcal{T}\Psi = (E - \mathcal{V})\Psi \tag{3.30b}$$

These two equations may be interpreted as follows. Equation (3.30a) shows that, in addition to or including those points in the configuration space where Ψ may vanish identically (see Section 3.2.1), there is a region defined by $\mathcal{V} = 0$ where the potential energy vanishes; this region is the same for every eigenstate of the system. Equation (3.30b) shows that, in addition to or including those points in the configuration space where Ψ may vanish identically, there is a region defined by $\mathcal{T}\Psi = (E-\mathcal{V})\Psi = 0$ where the kinetic energy vanishes; this region will be different for each eigenstate. These regions will be denoted hereafter as null potential-energy and null kinetic-energy regions (Fraga 1978, 1979a) and abbreviated as *NPER* and *NKER*, respectively. For example, for Helium-like atoms, with nuclear charge Z, the *NPER* and *NKER* will be defined (in atomic units) by

$$\frac{Z}{r_1} + \frac{Z}{r_2} - \frac{1}{r_{12}} = 0 \tag{3.31a}$$

$$E + \frac{Z}{r_1} + \frac{Z}{r_2} - \frac{1}{r_{12}} = 0 \tag{3.31b}$$

respectively.

Equation (3.31b) was studied in detail (Fraga 1978, 1979a) and the results applied to the prediction of lower bounds for the ground states of the He-isoelectronic series (Fraga 1981a) and generalized and applied to the semiempirical prediction of atomic ionization potentials (Fraga 1979b) and of atomic electron affinities (Fraga 1980) as well as to the study of the behaviour of atomic Hartree-Fock orbital energies (Fraga 1981b).

As was the case for the nodes and nodal planes, these regions are specific to the Schrödinger theory in the clamped-nuclei approximation.

3.3 Global Properties

We will now turn our attention to global properties of the eigenfunctions, expressed in terms of expectation values with integration over the whole configuration space, with particular emphasis on the molecular energy and its components.

3.3.1 The Potential Energy Hypersurface

The potential energy hypersurface embodies the dependence of the molecular energy on the parameters defining the nuclear geometry, as given by Eq. (3.7). It involves, therefore, the use of the clamped-nuclei Hamiltonian operator and it is in this connection that the problem arises. It is misleading just to invoke the fact that the nuclei are much heavier than the electrons and, in fact, care has to be exercised when removing the motion of the centre of mass.

We will now analyze the implications inherent to the use of the clamped-nuclei Hamiltonian operator and for that purpose we will follow the elegant work of Sutcliffe (1992, 1993, 1994, 1995). For simplicity, we will omit the mathematical development and refer the reader to the work of Sutcliffe as well as to related references [Eckart (1934, 1935), Hirschfelder and Wigner (1935), Berry (1960), Longuet-Higgins (1963), Kaplan (1975), Buck *et al.* (1979), Bunker (1979), Maruani and Serre (1983), Schmelzer and Murrell (1985), and Collins and Pearson (1993)].

The adoption of clamped nuclear positions represents, in fact, the adoption of a reference frame and two types of problems arise in this connection. On one hand, the clamped-nuclei Hamiltonian operator, viewed from the laboratory-fixed frame, is invariant under all the uniform translations and rigid rotations and rotations-reflections of the geometric figure specified by the nuclear positions and consequently its solutions will depend in a non-trivial form on the (3N-6) parameters that define the molecular geometry. On the other hand, the molecular Hamiltonian operator is invariant under a uniform translation and under a rigid rotation of the laboratory-fixed frame and it is necessary to separate the motion of the centre of mass, with a clear notational distinction between the nuclei and the electrons, and ensure the compatibility with the clamped-nuclei problem.

The procedure implies the removal of the translational motion, the distinction between nuclear and electronic motions, the adoption of body-fixed variables, and the separation of the nuclear and electronic motions. The matching of the electronic Hamiltonian operator obtained and the clamped-nuclei Hamiltonian operator requires that the solutions of the latter, that differ from one another only by nuclear translations or rotations, be excluded.

The adoption of the body-fixed variables is based on the fact that it is possible to make a transformation of the translation-free coordinates such that the rotational motion may be expressed in terms of three orientation variables and the remaining motions in terms of internal coordinates, which are invariant under all the orthogonal transformations of the translation-free coordinates. Given N nuclei, there are $\frac{1}{2}$N(N-1) internuclear distances, only (3N-6) of which may be independent. Therefore, for N ≥ 5, the number of internuclear distances is larger than the number of internal coordinates. That is, the set of internuclear distances is redundant and an independent subset must be chosen. The problem exists that it is possible to construct two or more geometric figures, for the molecular geometry under consideration, in which all the chosen internuclear distances are the same.

In the clamped-nuclei approach, the nuclei are identified and therefore the permutational symmetry of the problem is destroyed. The evaluation of the electronic energy must be carried out for a function that belongs to the totally symmetric representation of the overall permutation group of the identical nuclei and the properly invariant potential energy surface is then obtained by matching the behaviour of the nuclear repulsion energy.

In actual calculations, the invariance of the clamped-nuclei Hamiltonian operator is recognized and it is assumed that the electronic energy for a given function will have that value for all the functions that refer to the permuted nuclear variables.

We will assume then that the clamped-nuclei Hamiltonian operator is used and proceed to inspect additional global properties.

3.3.2 The Hellmann-Feynman Theorem

Let us consider the clamped-nuclei Hamiltonian operator (including, if appropriate, the nuclear repulsion energy term; see below). From the expectation value

$$<\Psi|(\mathcal{H} - E)\Psi> = 0 \tag{3.32}$$

one can see that $(\mathcal{H} - E)\Psi> = <\Psi|(\mathcal{H} - E) = 0$. Differentiating Eq. (3.32) with respect to a real parameter p in $\mathcal{H}$ yields

$$< \frac{\partial \Psi}{\partial p}|(\mathcal{H} - E)\Psi > + < \Psi|\frac{\partial(\mathcal{H} - E)}{\partial p} \Psi > + < \Psi|(\mathcal{H} - E)\frac{\partial \Psi}{\partial p} > = < \Psi|\frac{\partial(\mathcal{H} - E)}{\partial p} \Psi > = 0$$

This equation is usually rewritten as

$$\frac{\partial E}{\partial p} = \frac{< \Psi|\frac{\partial \mathcal{H}}{\partial p} \Psi >}{< \Psi|\Psi >} \tag{3.33}$$

which is the expression of the Hellmann-Feynmann theorem (Hellmann 1937, Feynman 1939). [See also Van Vleck (1928, 1932) and the review works of Epstein (1981) and Debb (1981).] A p-dependent transformation of the coordinate system will leave unchanged the numerical value of the *rhs* of Eq. (3.33) but will change its interpretation (Epstein 1981).

The so-called *electrostatic* Hellmann-Feynman theorem corresponds to the case when p is one of the nuclear position parameters, say a_k. The a_k-dependent part of the potential is

$$\mathcal{V}(a_k) = - \sum_i \frac{Z_k}{r_{ik}} + \sum_{\ell \neq k} \frac{Z_k Z_\ell}{r_{\ell k}}$$

and one obtains

$$F_x(\mathbf{a}_k) = -\frac{\partial \mathcal{V}(\mathbf{a}_k)}{\partial x_k} = \sum_i \frac{Z_k(x_i - x_k)}{r_{ik}^3} - \sum_{\ell \neq k} \frac{Z_k Z_\ell(x_\ell - x_k)}{r_{\ell k}^3} \qquad (3.34a)$$

$$F_y(\mathbf{a}_k) = -\frac{\partial \mathcal{V}(\mathbf{a}_k)}{\partial y_k} = \sum_i \frac{Z_k(y_i - y_k)}{r_{ik}^3} - \sum_{\ell \neq k} \frac{Z_k Z_\ell(y_\ell - y_k)}{r_{\ell k}^3} \qquad (3.34b)$$

$$F_z(\mathbf{a}_k) = -\frac{\partial \mathcal{V}(\mathbf{a}_k)}{\partial z_k} = \sum_i \frac{Z_k(z_i - z_k)}{r_{ik}^3} - \sum_{\ell \neq k} \frac{Z_k Z_\ell(z_\ell - z_k)}{r_{\ell k}^3} \qquad (3.34c)$$

and

$$\mathbf{F}(\mathbf{a}_k) = \mathbf{i}\,F_x(\mathbf{a}_k) + \mathbf{j}\,F_y(\mathbf{a}_k) + \mathbf{k}\,F_z(\mathbf{a}_k)$$

which is the force, due to all the electrons and the remaining nuclei, acting on the nucleus at position $\mathbf{a}_k$. Eq. (3.33) becomes then

$$\frac{\partial E}{\partial a_k} = -\frac{<\Psi|\mathbf{F}(\mathbf{a}_k)\Psi>}{<\Psi|\Psi>} \qquad (3.35)$$

which is the expression of the electrostatic Hellmann-Feynman theorem.

3.3.3 The Hypervirial Theorem

Given a linear operator $\mathcal{P}$ we can write

$$<\Psi|(\mathcal{HP} - \mathcal{PH})\Psi> = <\Psi|\mathcal{H}(\mathcal{P}\Psi)> - <\Psi|\mathcal{P}(\mathcal{H}\Psi)> = <\Psi|\mathcal{H}(\mathcal{P}\Psi)> - E<\Psi|\mathcal{P}\Psi>$$

which, taking into account the Hermitian character of $\mathcal{H}$, transforms into

$$<\Psi|\mathcal{H}(\mathcal{P}\Psi)> - E<\Psi|\mathcal{P}\Psi> = <\mathcal{H}\Psi|\mathcal{P}\Psi> - E<\Psi|\mathcal{P}\Psi>$$

$$= <\mathcal{P}\Psi|\mathcal{H}\Psi>^* - E<\Psi|\mathcal{P}\Psi> = E<\mathcal{P}\Psi|\Psi>^* - E<\Psi|\mathcal{P}\Psi>$$

$$= E<\Psi|\mathcal{P}\Psi> - E<\Psi|\mathcal{P}\Psi> = 0$$

That is: The expectation value

$$<\Psi|[\mathcal{H},\mathcal{P}]\Psi> = 0 \qquad (3.36)$$

of the commutator of $\mathcal{H}$ and $\mathcal{P}$ vanishes.

The hypervirial theorem is related to the Hellmann-Feynman theorem (Epstein 1965, Morgan and Landsberg 1965; see also Hirschfelder 1960, Coulson and Hurley 1962, Hirschfelder and Coulson 1962, Phillipson 1963, Hurley 1964, Benston 1965, Hirschfelder and Eliason 1967, and Steiner 1973). Given a unitary operator

$$\mathcal{U} = e^{ip\mathcal{P}} \qquad\qquad \mathcal{U}^{-1} = e^{-ip\mathcal{P}}$$

where p is a real parameter (not contained in the Hamiltonian operator) and $\mathcal{P}$ is a Hermitian operator, one can see first that

$$\langle\Psi|\mathcal{H}|\Psi\rangle = \langle\Psi|(\mathcal{U}^{-1}\mathcal{U})\mathcal{H}(\mathcal{U}^{-1}\mathcal{U})\Psi\rangle = \langle\mathcal{U}\Psi|(\mathcal{U}\mathcal{H}\mathcal{U}^{-1})(\mathcal{U}\Psi)\rangle$$

$$\langle\Psi|\Psi\rangle = \langle\Psi|(\mathcal{U}^{-1}\mathcal{U})\Psi\rangle = \langle\mathcal{U}\Psi|\mathcal{U}\Psi\rangle$$

That is: The equation

$$\langle\Psi|\mathcal{H}|\Psi\rangle = E\langle\Psi|\Psi\rangle$$

may be transformed into

$$\langle U\Psi|(U\mathcal{H}U^{-1})(U\Psi)\rangle = E\langle U\Psi|U\Psi\rangle$$

which shows that the function

$$\psi = e^{ip\mathcal{P}}\Psi$$

has, with respect to the transformed Hamiltonian operator

$$\mathcal{H}(p) = e^{ip\mathcal{P}}\ \mathcal{H}\ e^{-ip\mathcal{P}} \qquad\qquad (3.37)$$

the same eigenvalue E as the original eigenfunction with respect to $\mathcal{H}$. Therefore, application of the Hellmann-Feynman theorem to $\mathcal{H}(p)$, Eq. (3.36), will yield (taking into account that p is not contained in $\mathcal{H}$)

$$\frac{\partial E}{\partial p} = 0 = \frac{\langle\psi|\dfrac{\partial\mathcal{H}(p)}{\partial p}|\psi\rangle}{\langle\psi|\psi\rangle}$$

while from Eq. (3.37) we obtain

$$\frac{\partial\mathcal{H}(p)}{\partial p} = i\ e^{ip\mathcal{P}}\ (\mathcal{P}\mathcal{H} - \mathcal{H}\mathcal{P})\ e^{-ip\mathcal{P}}$$

so that

$$<\psi| \, e^{ipP}(P\mathcal{H} - \mathcal{H}P) \, e^{-ipP} \, \psi> = 0$$

Setting $p = 0$ one finally obtains

$$<\Psi|(P\mathcal{H} - \mathcal{H}P)\Psi> = <\Psi|[P,\mathcal{H}]\Psi> = 0$$

which is again the expression, Eq. (3.36), of the hypervirial theorem.

The hypervirial theorems have been treated exhaustively by Fernandez and Castro (1987) and therefore the reader interested in a complete analysis of the subject is referred to their work.

3.3.4 The Virial Theorem

Among other applications of the hypervirial theorem (Hirschfelder 1960, Epstein and Epstein 1962) it is worthwhile to mention that it can be used to obtain the virial theorem.

For an operator

$$P = \sum_i r_i \cdot p_i$$

and taking into account that

$$r_i[\mathcal{H},p_i] + [\mathcal{H},r_i]p_i = r_i(\mathcal{H}p_i - p_i\mathcal{H}) + (\mathcal{H}r_i - r_i\mathcal{H})p_i = \mathcal{H}(r_i \cdot p_i) - (r_i \cdot p_i)\mathcal{H}$$

$$= [\mathcal{H},r_i \cdot p_i]$$

the hypervirial theorem yields

$$<\Psi|[\mathcal{H},\sum_i r_i \cdot p_i]\Psi> = <\Psi|\{\sum_i r_i[\mathcal{H},p_i] + \sum_i [\mathcal{H},r_i]p_i\}\Psi> = 0 \qquad (3.38)$$

The relationships of interest for the evaluation of the commutators $[\mathcal{H},r_i]$ and $[\mathcal{H},p_i]$ are:

$$[p_i,r_i] = -i\hbar \qquad\qquad [p_i,r_j] = 0 \ (i \neq j)$$

$$[p_i \cdot p_i, r_i] = -2i\hbar \ p_i \qquad\qquad [p_i \cdot p_i, r_j] = 0 \ (i \neq j)$$

$$[p_i \cdot p_i, p_j] = 0 \qquad\qquad \text{(for any j)}$$

$$[\mathcal{V},r_i] = 0 \qquad\qquad [\mathcal{V},p_i] = -p_i\,\mathcal{V}$$

where $\mathcal{V}$ is a Coulomb potential. Therefore for a Hamiltonian operator, such as the one given by Eq. (3.2), one obtains

$$[\mathcal{H},r_i] = -\frac{i\hbar}{m_i}\,p_i \qquad\qquad [\mathcal{H},p_i] = -\,p_i\mathcal{V}$$

and Eq. (3.38) yields

$$< \Psi|\{\sum_i (r_i{\bullet}p_i\mathcal{V}) + i\hbar \sum_i \frac{1}{m_i}(p_i{\bullet}p_i)\}\Psi> \; = 0$$

which may be rewritten as

$$2i\hbar < \Psi|\mathcal{T}\Psi> \; = -< \Psi|\sum_i (r_i{\bullet}p_i\mathcal{V})\Psi>$$

or, equivalently,

$$2 < \Psi|\mathcal{T}\Psi> \; = \; < \Psi|\sum_i (r_i{\bullet}\nabla_i\mathcal{V})\Psi> \qquad\qquad (3.39)$$

which is the general expression of the virial theorem (see, e.g., the work of Hylleraas 1929, Fock 1930, and Slater 1933).

The *lhs* of this equation will always have the same form: that is, two times the expectation value of the kinetic energy for the particles considered in $\mathcal{P}$ and present in the Hamiltonian operator. The actual expression of the *rhs*, however, will depend on the case considered (atoms or molecules) and the composition of the potential operator, as discussed below:

(a) Atoms

With

$$\mathcal{V} = -Z\sum_i \frac{e^2}{r_{ia}} + \sum_{i<j} \frac{e^2}{r_{ij}} \qquad\qquad \mathcal{P} = \sum_i r_i{\bullet}p_i + r_a{\bullet}p_a$$

one obtains [see, e.g., Eqs. (3.34)]

$$\sum_i (r_i{\bullet}\nabla_i\mathcal{V}) + r_a{\bullet}\nabla_a\mathcal{V} = Z\sum_i \frac{e^2}{r_{ia}} - \sum_{i<j} \frac{e^2}{r_{ij}} = -\mathcal{V}$$

so that

$$2\langle\Psi|\mathcal{T}\Psi\rangle = -\langle\Psi|\mathcal{V}\Psi\rangle \tag{3.40}$$

[See also the derivation given by Bethe and Salpeter (1957).] Combining this equation with the expression for the total energy one obtains

$$E = \langle\Psi|\mathcal{T}\Psi\rangle + \langle\Psi|\mathcal{V}\Psi) = -\langle\Psi|\mathcal{T}\Psi\rangle = \tfrac{1}{2}\langle\Psi|\mathcal{V}\Psi\rangle$$

(b) Molecules

Within the context of the clamped-nuclei approximation, with

$$\mathcal{V} = -\sum_i \sum_k \frac{Z_k e^2}{r_{ik}} + \sum_{i<j} \frac{e^2}{r_{ij}} + \sum_{k<\ell} \frac{Z_k Z_\ell e^2}{r_{k\ell}} \qquad \mathcal{P} = \sum_i r_i \cdot p_i + \sum_k a_k \cdot p_k$$

and taking into account the electrostatic Hellmann-Feynman theorem, Eq. (3.35), for a normalized function, one obtains

$$2\langle\Psi|\mathcal{T}\Psi\rangle = -\langle\Psi|\mathcal{V}\Psi\rangle + \langle\Psi|\sum_k a_k \cdot F(a_K)\Psi\rangle \tag{3.41}$$

with the appearance of an additional term, when compared to the simple expression obtained for atoms. [See also the derivation of Levine (1991), using the properties of homogeneous functions and the Euler theorem, and the work of Parr and Brown (1968) and Nelander (1969).]

Regarding the practical use of the virial theorem as a criterion of the accuracy of a function, it is necessary to state the theorem obtained by Löwdin (1959) in his work on the scaling problem: *The fulfillment of the virial theorem is a necessary but not sufficient criterion that a wave function is an accurate solution of the Schrödinger equation corresponding to a stationary state.*

4 The Schrödinger Equation In Momentum Space

4.1 Introduction

Throughout the years most of the efforts dedicated to the Schrödinger equation in quantum and computational chemistry have been directed towards its (approximate) solution in position space (i.e., in the **r**-representation) and to the development of the concepts associated with, as well as the information to be obtained from, the corresponding functions $\Psi(\mathbf{r})$.

In particular, a detailed knowledge of the electron density distribution and of the electronic motion leads to an understanding of the molecular structure and of the chemical reactivity. The significance of the valence electrons in this regard, explored by Fukui (1952) within the framework of the frontier electron theory, is readily recognized at present, being accepted that the chemical reactivity is determined to a large extent by the electron density of the highest occupied molecular orbital.

It is in this connection that the study of the Schrödinger equation in momentum space (i.e., in the **p**-representation) has an additional interest as a source of new information, as outlined below. Work within the context of the **p**-representation, however, has lagged behind, except for some attempts in the early stages of Quantum Mechanics (Weyl 1928, Podolsky and Pauling 1929, Hylleraas 1932, and Fock 1935), but the development of new spectroscopic techniques has focused again the attention on the **p**-representation.

In this chapter, after a brief comparison of the advantages/handicaps of the two representations, we will present the fundamental formulations in the **p**-representation, which will be complemented with a survey of the trends in this field and illustrated with some representative numerical results.

4.1.1 The r-Representation versus the p-Representation

There are several reasons for the preferred interest in the functions $\Psi(\mathbf{r})$ and for the new work in the **p**-representation (Defranceschi and Delhalle 1990).

An important consideration is related to the interpretation and visualization of the results, which are difficult in the **p**-representation while leading rather naturally to the concept of electronic shells in the **r**-representation. An additional difficulty in the **p**-representation arises from the fact that the Schrödinger equation in that representation is an integral equation but, on the other hand, the singularities in $\mathcal{H}\Psi(\mathbf{r})$ (see Section 3.2.3) and the continuous character of the first derivatives of $\psi(\mathbf{p})$ are factors to be taken into consideration (Shibuya and Wulfman 1965, Navaza and Tsoucaris 1981).

The renewed efforts in the study of the **p**-representation, however, are due to the interest in the calculation of expectation values, associated with physical observables, as well as other momentum-related properties (Benesch and Smith 1973, Williams 1977), which can now be measured with modern experimental techniques (see, e.g., the work of Moore *et al.*, 1982, Brion 1986, McCarthy and Weigold 1991). Electron momentum spectroscopy (*EMS*), providing direct information on the electron momentum density distribution, has confirmed the importance, and consequently the need, of a knowledge of the density distribution in both the highest occupied (*HOMO*) and the lowest unoccupied (*LUMO*) molecular orbitals, as a source of more useful information than it may be obtained from the total density distribution (Allan and Cooper 1995).

The study of the Schrödinger equation in momentum space may proceed along two paths, which have in common the use of the Fourier Transform (*FT*) method. On one hand one can perform the transformation of the equation and try and solve it, which is the approach adopted by Fock (1935). Alternatively one can obtain directly the transforms of the functions $\Psi(\mathbf{r})$ and use the resulting functions $\psi(\mathbf{p})$ in the evaluation of whatever information is sought, as done, for example, by Podolsky and Pauling (1929).

It is therefore convenient, first of all, to outline the basic ideas of the *FT* method, which are needed for an understanding of the developments presented below, referring the reader to the work of Sneddon (1951), Champeney (1973), Defranceschi and Delhalle (1990), and Press *et al.* (1992) for additional details.

4.1.2 The Fourier Transform Method

The Fourier Transforms of two functions, F(u) and f(v), represent the back-and-forth transformation of a same function. The direct and inverse Fourier integral transforms are given, in general, by

$$F(u) = \int_{-\infty}^{\infty} f(v)\, e^{2\pi i u v}\, dv \qquad f(v) = \int_{-\infty}^{\infty} F(u)\, e^{-2\pi i u v}\, du \qquad (4.1)$$

Given two functions, f(v) and g(v), and their Fourier Transforms, F(u) and G(u), their convolution and correlation are defined by

$$C_n(f,g) = \int_{-\infty}^{\infty} f(w)\, g(v-w)dw \qquad C_r(f,g) = \int_{-\infty}^{\infty} f(w+v)\, g(w)dw \qquad (4.2)$$

respectively, with the properties that the FT of the convolution is equal to $F(u)G(u)$ and that the FT of the correlation is equal to $F(u)G^*(u)$.

If the function $f(v)$ has been sampled at a finite number of points, v_k, $k = 0,1,2,..., (N-1)$, with equal spacing Δ of the variable, so that $v_k = k\Delta$, then the discrete *FT* of $f(v)$ is given by

$$F(u_\ell) = \int_{-\infty}^{\infty} f(v)e^{2\pi i u_\ell v}\, dv \simeq \Delta \sum_{k=0}^{N-1} f_k e^{2\pi i u_\ell v_k} = \Delta \sum_{k=0}^{N-1} f_k e^{2\pi i k\ell/N} = \Delta F_\ell \qquad (4.3)$$

where

$$f_k = f(v_k) \qquad u_\ell = \frac{\ell}{N\Delta} \qquad \ell = -\frac{N}{2}, ..., \frac{N}{2} \qquad F_\ell = \sum_{k=0}^{N-1} f_k e^{2\pi i k\ell/N}$$

The discrete fast Fourier Transform (*FFT*) is based on the fact that the discrete *FT* of length N can be written as the sum of two discrete *FT*, each of length $N/2$, with the corresponding saving in computing time [see Section 5.1.1(b)].

In the present case, for the position-momentum transformation, the expressions customarily used are

$$\Psi(r) = (2\pi\hbar)^{-3/2} \int e^{i(\mathbf{p}\cdot\mathbf{r})/\hbar}\, \psi(\mathbf{p})d\mathbf{p} \qquad \psi(\mathbf{p}) = (2\pi\hbar)^{-3/2} \int e^{-i(\mathbf{p}\cdot\mathbf{r})/\hbar}\, \Psi(\mathbf{r})d\mathbf{r}$$

with integration over all space and with the factor $\hbar$ omitted when operating in atomic units. If $\Psi(\mathbf{r})$ is normalized we can write

$$\int \Psi^*(\mathbf{r})\Psi(\mathbf{r})d\mathbf{r} = (2\pi)^{-3} \int d\mathbf{r} \int d\mathbf{p} \int d\mathbf{p}'\, \psi^*(\mathbf{p}')\psi(\mathbf{p})\, e^{i(\mathbf{p}-\mathbf{p}')\cdot\mathbf{r}} = 1$$

so that, taking into account that

$$(2\pi)^{-3/2} \int e^{i(\mathbf{p}-\mathbf{p}')\cdot\mathbf{r}}d\mathbf{r} = \delta(\mathbf{p}-\mathbf{p}')$$

where $\delta(\mathbf{p}-\mathbf{p}')$ is the Kronecker delta, we obtain

$$\int d\mathbf{p} \int d\mathbf{p}'\, \psi^*(\mathbf{p}')\psi(\mathbf{p})\, \delta(\mathbf{p}-\mathbf{p}') = \int d\mathbf{p}\, \psi^*(\mathbf{p}')\psi(\mathbf{p}) = 1$$

which shows that normalization is preserved under the transformation.

For simplicity in the development of the formulation below we will summarize here the notation to be used (Defranceschi and Delhalle 1990). Starting with a function $f(\mathbf{r})$, its *FT* will be expressed as $f^T(\mathbf{p}) \equiv [f(\mathbf{r})]^T(\mathbf{p})$, so that the inverse FT will be given as $f(\mathbf{r}) = [f^T(\mathbf{p})]^T(\mathbf{r})$. The *FT* of translated functions are given by

$$[F(\mathbf{r}-\mathbf{r}')]^T(\mathbf{p}) = e^{-i(\mathbf{p}\cdot\mathbf{r}')} f^T(\mathbf{p}) \qquad [f^T(\mathbf{p}-\mathbf{p}')]^T(\mathbf{r}) = e^{i(\mathbf{p}'\cdot\mathbf{r})} f(\mathbf{r})$$

while for the convolution theorem we will write

$$[f(\mathbf{r})g(\mathbf{r})]^T(\mathbf{p}) = (2\pi)^{-3/2}\int d\mathbf{p}'\, f^T(\mathbf{p}')g^T(\mathbf{p}-\mathbf{p}')$$

$$f^T(\mathbf{p})g^T(\mathbf{p}) = [\int d\mathbf{r}'\, f(\mathbf{r}')g(\mathbf{r}-\mathbf{r}')]^T(\mathbf{p})$$

4.2 The Transformed Equation

Taking into account the purpose of this work, we will now proceed directly with the development of the formulation for the transformation of the Schrödinger equation, first in general and then for the particular case of the Hydrogen atom. For completeness, we will mention that the transformation of the Dirac equation has been discussed by Elsasser (1933), Rubinowicz (1948), Levy (1950), Bethe and Salpeter (1957), and Ishikawa *et al.* (1987, 1988).

4.2.1 General Formulation

We will consider the Hamiltonian operator, Eq. (3.6), in the clamped-nuclei approximation and write (in atomic units)

$$(2\pi)^{-3N/2}\int d\mathbf{r}\, e^{-i\sum_i(\mathbf{p}_i\cdot\mathbf{r}_i)}(\mathcal{H}-E)\Psi(\mathbf{r})$$

$$= (2\pi)^{-3N/2}\int d\mathbf{r}\, e^{-i\sum_i(\mathbf{p}_i\cdot\mathbf{r}_i)}\sum_i\{(\frac{1}{2}\mathbf{p}_i\cdot\mathbf{p}_i - E) - \sum_k\frac{Z_k}{r_{ik}} + \sum_{j>i}\frac{1}{r_{ij}}\}\Psi(\mathbf{r}) = 0$$

where the summations over i, j extend to the N-electrons in the system, the summation over k extends to the nuclei, and $\Psi(\mathbf{r})$ and $d\mathbf{r}$ stand for $\Psi(\mathbf{r}_1,\mathbf{r}_2,...,\mathbf{r}_N)$ and $d\mathbf{r}_1 d\mathbf{r}_2...d\mathbf{r}_N$, respectively.

The transformation of the three terms of this equation is carried out as follows (Shibuya and Wulfman 1965, Lasettre 1973, Judd 1975, Monkhorst and Jeziorski

1979, Navaza and Tsoucaris 1981, Defranceschi and Delhalle 1990). The transformation of the first term is immediate and one obtains

$$(2\pi)^{-3N/2} \int dr\, e^{-i\sum_i (\mathbf{p}_i \cdot \mathbf{r}_i)} \sum_i (\tfrac{1}{2}\, \mathbf{p}_i \cdot \mathbf{p}_i - E)\, \Psi(\mathbf{r})$$

$$= \sum_i \{(\tfrac{1}{2}\, \mathbf{p}_i \cdot \mathbf{p}_i - E)\, \{\prod_j (2\pi)^{-3/2} \int dr_j\, e^{-i\,(\mathbf{p}_j \cdot \mathbf{r}_j)}\Psi(\mathbf{r})\} \tag{4.4}$$

$$= \sum_i (\tfrac{1}{2}\, \mathbf{p}_i \cdot \mathbf{p}_i - E)\, \psi(\mathbf{p})$$

with

$$\psi(\mathbf{p}) = [\Psi(\mathbf{r})]^T(\mathbf{p}) = \psi(\mathbf{p}_1, \mathbf{p}_2, ..., \mathbf{p}_N) \tag{4.5}$$

The first step in the transformation of the Coulombic terms consists of finding the *FT* of $1/r_{ik}$ and $1/r_{ij}$. The solution of the problem is simplified (Judd 1975, Gradshteyn and Ryzhik 1980, Avery *et al.* 1996) by considering the general case of a potential of the form $\exp\{-\zeta r\}/r$. Thus, for the nuclear attraction term we will write

$$[e^{-\zeta_{ik}/r_{ik}}]^T(\mathbf{p}_i - \mathbf{p}') = (2\pi)^{-3/2} \int dr_i\, e^{-i(\mathbf{p}_i - \mathbf{p}')\cdot \mathbf{r}_i}(e^{-\zeta_{ik}/r_{ik}})$$

$$= (2\pi)^{-3/2}\, e^{-i(\mathbf{p}_i - \mathbf{p}')\cdot \mathbf{r}_k} \int dr_{ik}\, e^{-i(\mathbf{p}_i - \mathbf{p}')\cdot \mathbf{r}_{ik}}(e^{-\zeta_{ik}/r_{ik}})$$

$$= (2\pi)^{-3/2}\, e^{-i(\mathbf{p}_i - \mathbf{p}')\cdot \mathbf{r}_k} \int_0^\infty dr_{ik} r_{ik} e^{-\zeta_{ik}} \int_0^\pi d\theta_{ik}\, \sin\theta_{ik} \int_0^{2\pi} d\varphi_{ik} e^{-i(\mathbf{p}_i - \mathbf{p}')\cdot \mathbf{r}_{ik}}$$

with the centre of the polar coordinates on the nucleus k. Expanding (Edmonds 1960) the last exponential factor and taking $(\mathbf{p}_i - \mathbf{p}')$ in the direction of the polar axis we obtain

$$[e^{-\zeta_{ik}/r_{ik}}]^T(\mathbf{p}_i - \mathbf{p}') = (2\pi)^{-3/2}\, e^{-i(\mathbf{p}_i - \mathbf{p}')\cdot \mathbf{r}_k}$$

$$\sum_{\ell=0}^\infty (2\ell+1)(-i)^\ell \int_0^\infty dr_{ik} r_{ik} j_\ell(|\mathbf{p}_i - \mathbf{p}'|r_{ik}) e^{-\zeta_{ik}} \int_0^\pi d\theta_{ik}\, \sin\theta_{ik} P_\ell(\cos\theta_{ik}) \int_0^{2\pi} d\varphi_{ik}$$

$$= -(2\pi)^{-1/2}\, e^{-i(\mathbf{p}_i - \mathbf{p}')\cdot \mathbf{r}_k} \int_0^\infty dr_{ik} r_{ik} j_0(|\mathbf{p}_i - \mathbf{p}'|r_{ik}) e^{-\zeta_{ik}}$$

$$= (2/\pi)^{1/2} \frac{e^{-i(\mathbf{p}_i - \mathbf{p}')\cdot \mathbf{r}_k}}{(|\mathbf{p}_i - \mathbf{p}'|)^2 + \zeta^2}$$

where $P_\ell(\cos\theta_{ik})$ is a Legendre polynomial of the first kind (Jahnke and Emde 1945) and $j_\ell(|\mathbf{p}_i - \mathbf{p}'|r_{ik})$ is a Bessel function of the first kind (Abramowitz and Stegun 1970). Therefore, for the case under consideration we obtain

$$[1/r_{ik}]^T(|\mathbf{p}_i - \mathbf{p}'|) = (2/\pi)^{1/2} \frac{e^{-i(\mathbf{p}_i - \mathbf{p}')\cdot\mathbf{r}_k}}{(|\mathbf{p}_i - \mathbf{p}'|)^2}$$

$$1/r_{ik} = [[1/r_{ik}]^T(\mathbf{p}_i - \mathbf{p}')]^T(\mathbf{r}) = (2/\pi)^{-3/2} \int d\mathbf{p}'\, e^{i(\mathbf{p}_i - \mathbf{p}')\cdot\mathbf{r}_i} \{(2/\pi)^{1/2} \frac{e^{-i(\mathbf{p}_i - \mathbf{p}')\cdot\mathbf{r}_k}}{(|\mathbf{p}_i - \mathbf{p}'|)^2}\}$$

$$= \frac{1}{2\pi^2} \int \frac{d\mathbf{p}'}{(|\mathbf{p}_i - \mathbf{p}'|)^2}\, e^{i(\mathbf{p}_i - \mathbf{p}')\cdot\mathbf{r}_{ik}}$$

Therefore we can now write

$$[(1/r_{ik})\Psi(\mathbf{r})]^T(\mathbf{p}) = (2\pi)^{-3N/2} \int d\mathbf{r}\, e^{-i\sum_j (\mathbf{p}_j\cdot\mathbf{r}_j)} \frac{\Psi(\mathbf{r})}{r_{ik}}$$

$$= (2\pi)^{-3N/2} \int d\mathbf{r}\, e^{-i\sum_j (\mathbf{p}_j\cdot\mathbf{r}_j)} \{\frac{1}{2\pi^2} \int \frac{d\mathbf{p}'}{(|\mathbf{p}_i - \mathbf{p}'|)^2}\, e^{i(\mathbf{p}_i - \mathbf{p}')\cdot\mathbf{r}_{ik}}\}$$

$$= \frac{1}{2\pi^2} \int \frac{d\mathbf{p}'}{(|\mathbf{p}_i - \mathbf{p}'|)^2}\, e^{-i(\mathbf{p}_i - \mathbf{p}')\cdot\mathbf{r}_k} \{(2\pi)^{-3N/2} \int d\mathbf{r}\, e^{-i\{\sum_{j\neq i} (\mathbf{p}_j\cdot\mathbf{r}_j) + \mathbf{p}'\cdot\mathbf{r}_i\}}\Psi(\mathbf{r})\}$$

$$= \frac{1}{2\pi^2} \int \frac{d\mathbf{p}'}{(|\mathbf{p}_i - \mathbf{p}'|)^2}\, e^{-i(\mathbf{p}_i - \mathbf{p}')\cdot\mathbf{r}_k}\psi_i(\mathbf{p})$$

with $\psi_i(\mathbf{p}) = \psi(\mathbf{p}_1, \mathbf{p}_2, ..., \mathbf{p}', ..., \mathbf{p}_N)$, where the coordinate $\mathbf{p}_i$ has been replaced with $\mathbf{p}'$, and the complete nuclear attraction term is given by

$$[-\sum_i \sum_k (Z_k/r_{ik})\Psi(\mathbf{r})]^T(\mathbf{p})$$

$$= -\frac{1}{2\pi^2} \sum_i \sum_k Z_k \int \frac{d\mathbf{p}'}{(|\mathbf{p}_i - \mathbf{p}'|)^2}\, e^{-i(\mathbf{p}_i - \mathbf{p}')\cdot\mathbf{r}_k}\psi_i(\mathbf{p}) \qquad (4.6)$$

$$= -\frac{1}{2\pi^2} \sum_k Z_k \int \frac{d(\Delta\mathbf{p}')}{(\Delta\mathbf{p}')^2}\, e^{i(\Delta\mathbf{p}\cdot\mathbf{r}_k)}\psi_n(\mathbf{p})$$

with $\Delta\mathbf{p}' = \mathbf{p}' - \mathbf{p}_i$, $\Delta p' = |\Delta\mathbf{p}'|$, and

$$\psi_n(\mathbf{p}) = \sum_i \psi_i(\mathbf{p}) = \sum_i \psi_i(\mathbf{p}_1, \mathbf{p}_2, ..., \mathbf{p}', ..., \mathbf{p}_N) \qquad (4.7)$$

Proceeding in a similar manner, the transformation of the electron-electron repulsion term yields

$$[\frac{1}{2} \sum_{i \neq j} (1/r_{ij}) \Psi(\mathbf{r})]^{T}(\mathbf{p}) = \frac{1}{4\pi^2} \int \frac{d(\Delta \mathbf{p'})}{(\Delta \mathbf{p'})^2} \sum_{i \neq j} \{\psi_{ij}(\mathbf{p}) + \psi_{ji}(\mathbf{p})\}$$

$$= \frac{1}{4\pi^2} \int \frac{d(\Delta \mathbf{p'})}{(\Delta \mathbf{p'})^2} \psi_r(\mathbf{p})$$

(4.8)

with

$$\psi_r(\mathbf{p}) = \sum_{i \neq j} \{\psi_{ij}(\mathbf{p}) + \psi_{ji}(\mathbf{p})\}$$

$$\psi_{ij}(\mathbf{p}) = \psi_{ij}(\mathbf{p}_1, \mathbf{p}_2, ..., \mathbf{p'}, ..., \mathbf{p}_i + \mathbf{p}_j - \mathbf{p'}, ..., \mathbf{p}_N)$$

(4.9)

where the coordinates $\mathbf{p}_i$ and $\mathbf{p}_j$ have been replaced with $\mathbf{p'}$ and $\mathbf{p}_i + \mathbf{p}_j - \mathbf{p'}$, respectively.

The transformed Schrödinger equation, obtained from Eqs. (4.4), (4.6), and (4.8) is then

$$\sum_{i} (\frac{1}{2} \mathbf{p}_i \cdot \mathbf{p}_i - E)\psi(\mathbf{p}) - \frac{1}{2\pi^2} \sum_{k} Z_k \int \frac{d(\Delta \mathbf{p'})}{(\Delta \mathbf{p'})^2} \psi_n(\mathbf{p}) + \frac{1}{4\pi^2} \int \frac{d(\Delta \mathbf{p'})}{(\Delta \mathbf{p'})^2} \psi_r(\mathbf{p})$$

with $\psi(\mathbf{p})$, $\psi_n(\mathbf{p})$, and $\psi_r(\mathbf{p})$ defined by Eqs. (4.5), (4.7), and (4.9), respectively. The interpretation of these functions (Navaza and Tsoucaris 1981) is that the nuclear attraction results in a momentum transfer to each electron, with change in the total momentum, while the electron-electron repulsion produces a momentum transfer between each pair of electrons, without change in the total momentum. Navaza and Tsoucaris (1981) proposed an iterative procedure for the solution of this equation and one iteration was performed for a Helium-like atom but it was concluded that even a single iteration would be costly for a molecule.

A formulation, parallel to the one presented above, has been developed for the transformation of the Hartree-Fock equations and the reader is referred to the work of Novosadov and Pogonin (1982), Rodriguez and Ishikawa (1988), Bertier *et al.*, (1990), and Defranceschi and Delhalle (1990), among others.

4.2.2 The Hydrogen Atom

In the case of Hydrogenic atoms, the Schrödinger equation in position space (for bound states, with $E < 0$) is

$$\{\frac{1}{2\mu}\ p^2 + \mathcal{V}(\mathbf{r})\}\Psi(\mathbf{r}) = E\Psi(\mathbf{r})$$

(in atomic units, but with the reduced mass of the electron, μ, included explicitly). Multiplication on the left with $(2\pi)^{-3/2}\exp\{-i(\mathbf{p} \cdot \mathbf{r})\}$ and integration over all space yields (Levy 1950), after reordering,

$$(2\pi)^{-3/2}\int dr\ e^{-i(\mathbf{p} \cdot \mathbf{r})}\{\frac{1}{2\mu}\ p^2 - E) + \mathcal{V}(\mathbf{r})\}\Psi(\mathbf{r})$$

$$= (\frac{1}{2\mu}\ p^2 - E)\{(2\pi)^{-3/2}\int dre^{-i(\mathbf{p} \cdot \mathbf{r})}\Psi(\mathbf{r})\} + (2\pi)^{-3/2}\int dre^{-i(\mathbf{p} \cdot \mathbf{r})}\mathcal{V}(\mathbf{r})\Psi(\mathbf{r})$$

$$= (\frac{1}{2\mu}\ p^2 - E)\psi(\mathbf{p}) + (2\pi)^{-3/2}\int dre^{-i(\mathbf{p} \cdot \mathbf{r})}\mathcal{V}(\mathbf{r})\Psi(\mathbf{r})$$

Expressing $\Psi(\mathbf{r})$ as the FT of $\psi(\mathbf{p}')$, the second term of this equation may be written as

$$(2\pi)^{-3/2}\int dr\ e^{-i(\mathbf{p} \cdot \mathbf{r})}\ \mathcal{V}(\mathbf{r})\Psi(\mathbf{r})$$

$$= (2\pi)^{-3/2}\int dre^{-i(\mathbf{p} \cdot \mathbf{r})}\ \mathcal{V}(\mathbf{r})\{(2\pi)^{-3/2}\int dp'e^{i(\mathbf{p}' \cdot \mathbf{r})}\psi(\mathbf{p}')\}$$

$$= (2\pi)^{-3}\int dp'\psi(\mathbf{p}')\{\int dre^{-i(\mathbf{p}-\mathbf{p}') \cdot \mathbf{r}}\ \mathcal{V}(\mathbf{r})\}$$

$$= \int dp'\mathcal{U}(\mathbf{p} - \mathbf{p}')\psi(\mathbf{p}')$$

with the *FT* of the potential $\mathcal{V}(\mathbf{r})$ given by

$$\mathcal{U}(\mathbf{p} - \mathbf{p}') = (2\pi)^{-3}\int dr\ e^{-i(\mathbf{p}-\mathbf{p}') \cdot \mathbf{r}}\ \mathcal{V}(\mathbf{r})$$

Therefore the transformed Schrödinger equation will be written as

$$(\frac{1}{2\mu}\ p^2 - E)\psi(\mathbf{p}) = -\int dp'\ \mathcal{U}(\mathbf{p} - \mathbf{p}')\psi(\mathbf{p}') \qquad (4.10)$$

For an isolated Hydrogenic atom, in which case $\mathcal{V}(\mathbf{r})$ is a Hermitian and real central potential, $\mathcal{U}(\mathbf{p}-\mathbf{p}')$ is given by

$$\mathcal{U}(\mathbf{p}-\mathbf{p'}) = (2\pi)^{-3} \int_0^\infty dr\, r^2 \mathcal{V}(r) \int_0^\pi d\theta \sin\theta \int_0^{2\pi} d\varphi\, e^{-i(\mathbf{p}-\mathbf{p'})\cdot\mathbf{r}}$$

$$= (2\pi)^{-2} \int_0^\infty dr\, r^2 \mathcal{V}(r) \int_0^\pi d\theta\, e^{-i|\mathbf{p}-\mathbf{p'}|r\cos\theta} \sin\theta$$

$$= \frac{1}{2\pi^2 |\mathbf{p}-\mathbf{p'}|} \int_0^\infty dr\, r \sin(|\mathbf{p}-\mathbf{p'}|r) \mathcal{V}(r)$$

with the integration performed in polar spherical coordinates, with the polar axis directed along $(\mathbf{p}-\mathbf{p'})$. The potential $\mathcal{U}(\mathbf{p}-\mathbf{p'})$ depends in this case on the absolute value $|\mathbf{p}-\mathbf{p'}|$. In addition, taking into account that $\mathcal{V}(r)$ is real and the fact that for two functions, $f^*(v) = g(v)$, their *FT* are $F^*(u) = G(-u)$, one concludes that $\mathcal{U}(\mathbf{p}-\mathbf{p'}) = \mathcal{U}^*(\mathbf{p}-\mathbf{p'})$.

In the case when $\mathcal{U}(\mathbf{p})$ is a central potential, Eq. (4.10) is separable in polar spherical coordinates, as it is the case for the Schrödinger equation in position space, when $\mathcal{V}(r)$ is a central potential. Using

$$\Psi(\mathbf{r}) = R_{n\ell}(r) Y_{\ell m}(\theta,\varphi) \tag{4.11}$$

where $R_{n\ell}(r)$ is the normalized radial function, $Y_{\ell m}(\theta,\varphi)$ is a normalized spherical harmonic, and (n, ℓ, m) are the usual quantum numbers, one obtains

$$\psi(\mathbf{p}) = (2\pi)^{-3/2} \int d\mathbf{r}\, e^{-i(\mathbf{p}\cdot\mathbf{r})} \Psi(\mathbf{r}) = (2\pi)^{-3/2} \int d\mathbf{r}\, e^{-i(\mathbf{p}\cdot\mathbf{r})} R_{n\ell}(r) Y_{\ell m}(\theta,\varphi)$$

$$= (2\pi)^{-3/2} \sum_{\ell'=0}^\infty (-i)^{\ell'} (2\ell'+1) \int_0^\infty dr\, r^2 j_{\ell'}(pr) R_{n\ell}(r)$$

$$\int_0^\pi d\theta \sin\theta\, P_{\ell'}(\cos\omega) \int_0^\pi d\varphi\, Y_{\ell m}(\theta,\varphi)$$

after expansion (see Section 4.2.1) of the exponential factor and where ω denotes the angle between the vectors $\mathbf{r}$ and $\mathbf{p}$. Using the addition theorem for the spherical harmonics as well as their orthonormality properties (Steinborn and Ruedenberg 1973) yields

$$\psi(\mathbf{p}) = (2/\pi)^{1/2} \sum_{\ell'=0}^\infty \sum_{m'=-\ell'}^{\ell'} (-i)^{\ell'} \int_0^\infty dr\, r^2 j_{\ell'}(pr) R_{n\ell}(r) \int_0^\pi d\theta \sin\theta$$

$$\int_0^{2\pi} d\varphi\, Y_{\ell' m'}^*(\theta,\varphi) Y_{\ell' m'}(\theta',\varphi') Y_{\ell m}(\theta,\varphi)$$

$$= (2/\pi)^{1/2} (-i)^\ell Y_{\ell m}(\theta',\varphi') \int_0^\infty dr\, r^2 j_\ell(pr) R_{n\ell}(r)$$

where (θ', φ') are the polar angles associated with $\mathbf{p}$. This equation, which may be used for the evaluation of the *FT* of Hydrogenic orbitals (Komarov and Temkin 1976, Kaijser and Smith 1977) may be rewritten as

$$\psi(\mathbf{p}) = F_{n\ell}(p)Y_{\ell m}(\theta,\varphi) \tag{4.12a}$$

with

$$F_{n\ell}(p) = (2/\pi)^{1/2}(-i)^{\ell}\int_0^{\infty} dr\, r^2 j_{\ell}(pr)\, R_{n\ell}(r) \tag{4.12b}$$

and having omitted the primes in the polar angles for convenience in the following formulation.

The final form of the transformed equation, in the present case, will be obtained (Levy 1950) by substitution of Eq. (4.12a) into Eq. (4.10) and subsequent manipulation. We will then write

$$(\frac{1}{2\mu}p^2 - E_n)F_{n\ell}(p)Y_{\ell m}(\theta,\varphi) = -\int d\mathbf{p}'\, \mathcal{U}(|\mathbf{p} - \mathbf{p}'|)F_{n\ell}(p')Y_{\ell m}(\theta',\varphi')$$

$$= -\int_0^{\infty} dp'(p')^2 F_{n\ell}(p') \int_0^{\pi} d\theta'\sin\theta' \int_0^{2\pi} d\varphi'\, Y_{\ell m}(\theta',\varphi')\mathcal{U}(|\mathbf{p} - \mathbf{p}'|) \tag{4.13}$$

$$= -\int_0^{\infty} dp'(p')^2 F_{n\ell}(p') \int_{-1}^{1} d(\cos\theta') \int_0^{2\pi} d\varphi'\, Y_{\ell m}(\theta',\varphi')\mathcal{U}(|\mathbf{p} - \mathbf{p}'|)$$

where (θ,φ) and (θ',φ') are the polar angles associated with $\mathbf{p}$ and $\mathbf{p}'$, respectively; $|\mathbf{p}-\mathbf{p}'|$ is given by

$$|\mathbf{p}-\mathbf{p}'|^2 = \begin{cases} p^2 + (p')^2 - 2pp'\cos(|\theta-\theta'|) & \text{in general} \\[2mm] p^2 + (p')^2 - 2pp'\cos(|\theta'|) & \text{with } \mathbf{p} \text{ in the direction of the polar axis} \end{cases}$$

and $\mathcal{U}(|\mathbf{p}-\mathbf{p}'|)$ will be noted, correspondingly, as $\mathcal{U}(p,p';\theta,\theta')$ and $\mathcal{U}(p,p';\theta')$.

The *rhs* of Eq. (4.13) may be simplified as follows (Salpeter 1951). We first obtain

$$\int_{-1}^{1} d(\cos\theta') \int_0^{2\pi} d\varphi'\, Y_{\ell m}(\gamma')P_{\ell}(\cos|\gamma-\gamma'|)$$

$$= \int_{-1}^{1} d(\cos\theta') \int_0^{2\pi} d\varphi'\, Y_{\ell m}(\gamma')\{\frac{4\pi}{2\ell+1}\sum_{m'=-\ell}^{\ell} Y_{\ell m'}^*(\gamma')Y_{\ell m'}(\gamma)\}$$

$$= \frac{4\pi}{2\ell+1} Y_{\ell m}(\gamma)$$

where γ and γ' denote the unit vectors (θ,φ) and (θ',φ'), respectively. We now apply this equation to the case

$$\int_{-1}^{1} d(\cos\theta') \int_{0}^{2\pi} d\varphi'\, Y_{\ell m}(\gamma\!+\!\gamma_0)P_\ell(\cos|\gamma|)$$

$$= \int_{-1}^{1} d(\cos\theta') \int_{0}^{2\pi} d\varphi'\, Y_{\ell m}(\gamma\!+\!\gamma_0)P_\ell(\cos|(\gamma+\gamma_0)-\gamma_0|) = \frac{4\pi}{2\ell+1}\, Y_{\ell m}(\gamma_0) \qquad (4.14)$$

Expressing $Y_{\ell m}(\gamma\!+\!\gamma_0)$ as

$$Y_{\ell m}(\gamma+\gamma_0) = \sum_{m'=-\ell}^{\ell} a_{m'}(\gamma_0)Y_{\ell m'}(\gamma)$$

we can write

$$\int_{-1}^{1} d(\cos\theta) \int_{0}^{2\pi} d\varphi\, Y_{\ell m}(\gamma+\gamma_0)P_\ell(\cos|\gamma|)$$

$$= \int_{-1}^{1} d(\cos\theta) \int_{0}^{2\pi} d\varphi\Big\{ \sum_{m'=-\ell}^{\ell} a_{m'}(\gamma_0)Y_{\ell m'}(\gamma) \Big\} P_\ell(\cos|\gamma|)$$

$$= \sum_{m'=-\ell}^{\ell} a_{m'}(\gamma_0) \int_{-1}^{1} d(\cos\theta)P_{\ell m'|}(\cos|\gamma|)P_\ell(\cos|\gamma|) \int_{0}^{2\pi} d\varphi\, e^{im'\varphi}$$

$$= \frac{4\pi}{2\ell+1}a_0(\gamma_0)$$

where $P_{\ell m'|}$ is a Legendre associated function of the first kind. Comparison of this result with the one obtained in Eq. (4.14) yields

$$a_0(\gamma_0) = Y_{\ell m}(\gamma_0)$$

and we can write

$$Y_{\ell m}(\gamma+\gamma_0) = a_0(\gamma_0)Y_{\ell 0}(\gamma) = Y_{\ell m}(\gamma_0)P_\ell(\cos|\gamma|)$$

In order to use this result in Eq. (4.13) it is necessary to adapt first the notation. If we take γ_0 as the polar axis and p parallel to γ_0, then $(\theta,\varphi)\equiv(0,0)$ and $(\gamma+\gamma_0)\equiv(\theta',\varphi')$, so that

$$Y_{\ell m}(\gamma+\gamma_0) = Y_{\ell m}(\gamma_0)\,P_\ell(\cos\theta')$$

where the notation $Y_{\ell m}(\gamma_0)$ has been maintained for simplicity. Substitution into Eq. (4.13) yields finally

$$(\frac{1}{2\mu} p^2 - E_n)F_{n\ell}(p)Y_{\ell m}(\gamma_0)$$

$$= - 2\pi \, Y_{\ell m}(\gamma_0)\int_0^\infty dp'(p')^2 F_{n\ell}(p') \int_{-1}^{1} d(\cos \theta')P_\ell(\cos \theta')\mathcal{U}(p,p';\theta')$$

which reduces to

$$(\frac{1}{2\mu} p^2 - E_n)F_{n\ell}(p) = - \int_0^\infty dp'(p')^2 K_\ell(p,p') \, F_{n\ell}(p') \qquad (4.15a)$$

where

$$K_\ell(p,p') = 2\pi \int_{-1}^{1} d(\cos \theta')P_\ell(\cos \theta')\mathcal{U}(p,p';\theta') \qquad (4.15b)$$

is a symmetric kernel in p and p' which depends on the value of ℓ and on the potential $\mathcal{V}(\mathbf{r})$ (see, e.g., Levy 1950 and Bethe and Salpeter 1957).

Equations (4.15) can be solved exactly for some simple potentials $\mathcal{V}(\mathbf{r})$ (Levy 1950) and here we will examine the case of the Coulomb potential, -Z/r. The *FT* of this potential may be obtained using the approach presented in Section 4.2.1 for a general potential. One obtains

$$\mathcal{U}(|\mathbf{p} - \mathbf{p'}|) = - \frac{Z}{2\pi^2(|\mathbf{p} - \mathbf{p'}|)^2}$$

so that

$$K_\ell(p,p') = - \frac{Z}{\pi} \int_{-1}^{1} \frac{1}{[p^2+(p')^2 - 2pp' \cos \theta \,']} P_\ell(\cos \theta \,')d(\cos \theta \,') = - \frac{Z}{\pi pp'} Q_\ell\{\frac{p^2+(p')^2}{2pp'}\}$$

where Q_ℓ is a Legendre polynomial of the second kind (Jahnke and Emde 1945). Substitution into Eq. (4.15a) yields finally the one-dimensional Schrödinger equation in momentum space

$$(\frac{1}{2\mu} p^2 - E_n)F_{n\ell}(p) = \frac{Z}{\pi p} \int_0^\infty dp' \, p' \, Q_\ell\{\frac{p^2+(p')^2}{2pp'}\}F_{n\ell}(p') \qquad (4.16)$$

which yields the same discrete eigenvalues as the Schrödinger equation in position space (Fock 1935). [The case of the continuous spectrum is discussed by Bethe and Salpeter (1957).]

4.3 The Transformed Functions

We have seen in the preceding section that the solution of the transformed Schrödinger equation offers serious computational difficulties, except for one-electron systems, and therefore one should consider the alternate approach of the direct transformation of the functions from the **r**- to the **p**-representation.

The availability of functions in the **p**-representation would allow us to study their properties as well as to calculate those expectation values of possible interest and it is in this connection that some comments are appropriate.

Exact analytical functions in the **r**-representation are only known for one-electron systems (Bates *et al.* 1953) while for all other atomic and molecular systems only computational functions (i.e., functions determined computationally) are available. Generally those functions are determined at an approximate level and only in a few cases quasi-exact or exact solutions [such as those of Pekeris (1958) and Kolos and Wolniewicz (1966)] have been determined.

As a consequence, the attention has been focused on the transformation of hydrogenic orbitals and of functions obtained in self-consistent field (*SCF*) calculations, within the context of the independent-particle and expansion approximations. Here we will discuss in detail the transformation of the hydrogenic orbitals but we will not consider the transformation of either *SCF* or Hartree-Fock functions, as they are outside the scope of this work. However, because of their illustrative interest and in the absence of results obtained from functions $\psi(\mathbf{p})$ determined directly, we will present below some examples of density distributions and expectation values obtained from functions in the **p**-representation generated from *SCF* functions in the **r**-representation.

4.3.1 Hydrogenic Orbitals

The formulation for the transformation of Hydrogenic orbitals was developed by Podolsky and Pauling (1929) and later simplified and complemented by Hey (1993) and Hoang Binh and Van Regemorter (1997). [See also Bransden and Joachain (1983).]

As already indicated in Section 4.2.2, the Hydrogenic orbitals in the **p**-representation are separable, Eq. (4.12a), and their radial components may be obtained by application of Eq. (4.12b). The radial functions in the **r**-representation are given by

$$R_{n\ell}(r) = \frac{[\Gamma(n-\ell)]^{1/2}}{n[\Gamma(n+\ell+1)]^2} r^{-1} e^{-r/n} \left(\frac{2r}{n}\right)^{\ell+1} L_{n+\ell}^{2\ell+1}\left(\frac{2r}{n}\right)$$

in terms of the Γ-function and the Laguerre polynomials $L_{n+\ell}^{2\ell+1}\left(\dfrac{2r}{n}\right)$ and therefore we can write

$$F_{n\ell}(p) = (2/\pi)^{1/2}(-i)^\ell \int_0^\infty dr\, r^2\, j_\ell(pr)\, R_{n\ell}(r)$$

$$= p^{-1/2}(-i)^\ell \int_0^\infty dr\, r^{3/2}\, J_{\ell+1/2}(pr)\, R_{n\ell}(r)$$

$$= p^{-1/2}(-i)^\ell \frac{[\Gamma(n-\ell)]^{1/2}}{n[\Gamma(n+\ell+1)]^{3/2}} \int_0^\infty dr\, r^{1/2}\, J_{\ell+1/2}(pr)\, e^{-r/n} \left(\frac{2r}{n}\right)^{\ell+1} L_{n+\ell}^{2\ell+1}\left(\frac{2r}{n}\right)$$

$$= p^{-1/2} A_{n\ell} \int_0^\infty dx\, x^{\ell+3/2}\, e^{-x/2}\, J_{\ell+1/2}(\beta x)\, L_{n+\ell}^{2\ell+1}(x)$$

$$= p^{-1/2} A_{n\ell} \int_0^\infty dx\, x^{\ell+3/2}\, e^{-x/2}\, J_{\ell+1/2}(\beta x)\, L_{2\ell+1+k}^{2\ell+1}(x)$$

$$= p^{-1/2} A_{n\ell}\, I_{\ell k}$$

$$(4.17)$$

with

$$A_{n\ell} = (-i)^\ell \frac{[n\Gamma(n-\ell)]^{1/2}}{2\sqrt{2}\,[\Gamma(n+\ell+1)]^{3/2}} \qquad \beta = \frac{1}{2}\, pn \qquad x = \frac{2r}{n} \qquad k = n-\ell-1$$

and where $J_{\ell+1/2}(\beta x)$ is a spherical Bessel function (Erdelyi *et al.* 1954, Abramowitz and Stegun 1970).

The solution of the problem is obtained (Podolsky and Pauling 1929, Hoang Binh and Van Regemorter 1997) through the use of the generating functions

$$(-1)^{2\ell+1} \frac{e^{-xy(1-y)}}{(1-y)^{2\ell+2}} = \sum_{k=0}^\infty \frac{y^k}{\Gamma(2\ell+2+k)} L_{2\ell+1+k}^{2\ell+1}(x)$$

$$\frac{1-y^2}{[1-2xy+y^2]^{\ell+2}} = \frac{1}{\ell+1} \sum_{k=0}^\infty (\ell+1+k)\, C_k^{\ell+1}(x)\, y^k$$

of the Laguerre polynomials and of the Gegenbauer polynomials, $C_k^{\ell+1}(x)$, respectively, *via* an expression which will allow us to establish a correspondence between the two summations. Such an expression is

$$\sum_{k=0}^\infty \frac{y^k}{\Gamma(2\ell+2+k)} I_{\ell k} = \int_0^\infty dx\, x^{\ell+3/2}\, e^{-x/2}\, J_{\ell+1/2}(\beta x) \sum_{k=0}^\infty \frac{y^k}{\Gamma(2\ell+2+k)} L_{2\ell+1+k}^{2\ell+1}(x)$$

$$= \frac{(-1)^{2\ell+1}}{(1-y)^{2\ell+2}} \int_0^\infty dx\, x^{\ell+3/2}\, e^{-x(1+y)/2(1-y)}\, J_{\ell+1/2}(\beta x)$$

which transforms, using the solution obtained by Watson (1966) for the integral, into

$$\sum_{k=0}^{\infty} \frac{y^k}{\Gamma(2\ell+2+k)} I_{\alpha k} = \frac{(-1)^{2\ell+1} 2^{2\ell+4} \Gamma(\ell+2)(np)^{\ell+1/2}}{\sqrt{\pi}(n^2 p^2 + 1)^{\ell+2}} \frac{1-y^2}{(1-2yz+y^2)^{\ell+2}}$$

$$= \frac{(-1)^{2\ell+1} 2^{2\ell+4} \Gamma(2\ell+2)(np)^{\ell+1/2}}{\sqrt{\pi}(n^2 p^2 + 1)^{\ell+2}(\ell+1)} \sum_{k=0}^{\infty} (\ell+1+k) C_k^{\ell+1}(z) y^k \tag{4.18}$$

where

$$z = (n^2 p^2 - 1)/(n^2 p^2 + 1)$$

From the correspondence between the terms of the two summations one obtains (Bransden and Joachain 1983, Hey 1993), for $k = n - \ell - 1$,

$$F_{n\ell}(p) = (-1)^{2\ell+1}(-i)^{\ell}(2/\pi)^{1/2} 2^{2\ell+2} \Gamma(\ell+1) \tag{4.19}$$

$$\left[\frac{\Gamma(n-\ell)}{\Gamma(n+\ell+1)} \right]^{1/2} \frac{n^2(np)^\ell}{(n^2 p^2 + 1)^{\ell+2}} C_{n-\ell-1}^{\ell+1} \left(\frac{n^2 p^2 - 1}{n^2 p^2 + 1} \right)$$

The Gegenbauer polynomials may be expressed in terms of Γ-functions (Rainville 1960) and may also be generated by a generalization of the Rodrigues formula for the Legendre polynomials (Magnus *et al.* 1966). They were tabulated by Podolsky and Pauling (1929) for the usual values of the quantum number ($n \leq 6$, $\ell \leq 4$).

These radial functions, which have the same number of nodes as the functions in the r-representation, vary as p^ℓ for small p (and therefore vanish at the origin, except for $\ell = 0$) and as $p^{-(\ell+4)}$ for large p, and they have the additional characteristic that their expectation values $<p^m>$ exist only for a limited number of integral values of m (Hey 1993).

4.4 Properties and Expectation Values

Taking into account the comments made at the beginning of the preceding section, regarding the lack of exact functions in the r-representation, the discussion in this section would have to be presented in a rather general way. For completeness, however, we will also include some mention of the work done with transformed functions.

4.4.1 Symmetry

The functions $\Psi(\mathbf{r})$ and $\psi(\mathbf{p})$, related as *FT* of each other, constitute different representations of the same vector in a Hilbert space and therefore they will have the same eigenvalue for all operators, including the operators of the symmetry point group defined by the nuclear geometry of the system (Berthier *et al.* 1989). [See Section 3.1.2(e).]

The result of the transformation by a unitary symmetry operator will be examined below at the orbital level (Defranceschi and Berthier 1990). We will write

$$\phi^{(S)}(\mathbf{r}) = \phi(\mathbf{r}^{(S)}) = \phi(\mathbf{S}^{-1}\mathbf{r})$$

where $\phi^{(S)}$ and $\mathbf{r}^{(S)}$ denote the orbital and position vector transformed under the symmetry operation and $\mathbf{S}^{-1}$ is the unitary matrix that effects that transformation of the position vector; $\mathbf{r}$ and $\mathbf{r}^{(S)}$ are column vectors and $\mathbf{S}^{-1}\mathbf{r}$ denotes a matrix product. The scalar product $\mathbf{p} \cdot \mathbf{r}^{(S)}$, where $\mathbf{p}$ is a row vector, may be written as

$$\mathbf{p} \cdot \mathbf{r}^{(S)} = \mathbf{p}\mathbf{S}^{-1}\mathbf{r} = \mathbf{r}^{\dagger}(\mathbf{S}^{-1})^{\dagger}\mathbf{p}^{\dagger} = \mathbf{r}^{\dagger}\mathbf{S}\mathbf{p}^{\dagger} = \mathbf{r}^{\dagger} \cdot \mathbf{p}^{(S)\dagger} = \mathbf{p}^{(S)} \cdot \mathbf{r}$$

where the expressions without the scalar product dot involve matrix operations and $\mathbf{S} = (\mathbf{S}^{-1})^{\dagger}$. Therefore the *FT* of $\phi(\mathbf{r}^{(S)})$ will be

$$[\phi(\mathbf{r}^{(S)})]^{T}(\mathbf{p}) = (2\pi)^{-3/2} \int d\mathbf{r}^{(S)} \, e^{-i(\mathbf{p} \cdot \mathbf{r}^{(S)})} \, \phi(\mathbf{r}^{(S)})$$

$$= (2\pi)^{-3/2} \int d\mathbf{r} \, e^{-i(\mathbf{p}^{(S)} \cdot \mathbf{r})} \, \phi(\mathbf{S}^{-1}\mathbf{r})$$

with $d\mathbf{r}^{(S)} \equiv d\mathbf{r}$. By analogy with the definition of $\phi(\mathbf{r}^{(S)})$ we can then write

$$[\phi(\mathbf{r}^{(S)})]^{T}(\mathbf{p}) = \chi(\mathbf{p}^{(S)}) = \chi(\mathbf{S}\mathbf{p})$$

which shows that if in the $\mathbf{r}$-representation there exists an orbital of a given symmetry under a given operator, then an orbital $\chi(\mathbf{p})$ of the same symmetry under the same operator will correspond to it in the $\mathbf{p}$-representation.

An interesting point, which has been studied by Defranceschi and Berthier (1990) at the orbital level, is the appearance of an inversion centre in the transformation from the $\mathbf{r}$- to the $\mathbf{p}$-representation. In the general case when $\phi(\mathbf{r})$ consists of both a real and an imaginary component, which we will denote by $\phi_{(R)}(\mathbf{r})$ and $\phi_{(I)}(\mathbf{r})$, respectively, we can write

$$\chi(\mathbf{p}) = (2\pi)^{-3/2} \int d\mathbf{r} \, e^{-i(\mathbf{p} \cdot \mathbf{r})} \{\phi_{(R)}(\mathbf{r}) + i \, \phi_{(I)}(\mathbf{r})\}$$

$$= (2\pi)^{-3/2} \int d\mathbf{r} \, [\cos(\mathbf{p} \cdot \mathbf{r}) - i \sin(\mathbf{p} \cdot \mathbf{r})] \{\phi_{(R)}(\mathbf{r}) + i \, \phi_{(I)}(\mathbf{r})\}$$

$$= \chi_{(R)}(\mathbf{p}) + i \, \chi_{(I)}(\mathbf{p})$$

where

$$\chi_{(R)}(\mathbf{p}) = (2\pi)^{-3/2} \int d\mathbf{r} \, \{\phi_{(R)}(\mathbf{r}) \cos(\mathbf{p} \cdot \mathbf{r}) + \phi_{(I)}(\mathbf{r}) \sin(\mathbf{p} \cdot \mathbf{r})\} \tag{4.20a}$$

$$\chi_{(I)}(\mathbf{p}) = (2\pi)^{-3/2} \int d\mathbf{r} \, \{-\phi_{(R)}(\mathbf{r}) \sin(\mathbf{p} \cdot \mathbf{r}) + \phi_{(I)}(\mathbf{r}) \cos(\mathbf{p} \cdot \mathbf{r})\} \tag{4.20b}$$

are the real and imaginary components of $\chi(\mathbf{p})$. For the complex conjugate of $\chi(\mathbf{p})$ we can then write

$$\chi^*(\mathbf{p}) = \chi_{(R)}(\mathbf{p}) - i \, \chi_{(I)}(\mathbf{p}) \tag{4.21}$$

or, alternatively,

$$\chi^*(\mathbf{p}) = ([\phi(\mathbf{r})]^T(\mathbf{p}))^* = (2\pi)^{-3/2} \left\{ \int d\mathbf{r} \, e^{-i(\mathbf{p} \cdot \mathbf{r})} \phi(\mathbf{r}) \right\}^*$$

$$= (2\pi)^{-3/2} \int d\mathbf{r} \, e^{-i(-\mathbf{p}) \cdot \mathbf{r}} \phi^*(\mathbf{r}) = [\phi^*(\mathbf{r})]_{(R)}^T(-\mathbf{p}) + i \, [\phi^*(\mathbf{r})]_{(I)}^T(-\mathbf{p})$$

so that, taking into account Eqs. (4.20) and (4.21), we obtain

$$\chi_{(R)}(\mathbf{p}) = (2\pi)^{-3/2} \left\{ \int d\mathbf{r} \, \{-\phi_{(R)}(\mathbf{r}) \cos(\mathbf{p} \cdot \mathbf{r}) + \phi_{(I)}(\mathbf{r}) \sin(\mathbf{p} \cdot \mathbf{r})\} = [\phi^*(\mathbf{r})]_{(R)}^T(-\mathbf{p}) \right.$$

$$\chi_{(I)}(\mathbf{p}) = (2\pi)^{-3/2} \left\{ \int d\mathbf{r} \, \{-\phi_{(R)}(\mathbf{r}) \sin(\mathbf{p} \cdot \mathbf{r}) + \phi_{(I)}(\mathbf{r}) \cos(\mathbf{p} \cdot \mathbf{r})\} = -[\phi^*(\mathbf{r})]_{(I)}^T(-\mathbf{p}) \right.$$

Therefore, if $\phi(\mathbf{r})$ is real we will have

$$\chi_{(R)}(\mathbf{p}) = (2\pi)^{-3/2} \int d\mathbf{r} \, \phi(\mathbf{r}) \cos(\mathbf{p} \cdot \mathbf{r}) = [\phi(\mathbf{r})]_{(R)}^T(-\mathbf{p})$$

$$\chi_{(I)}(\mathbf{p}) = -(2\pi)^{-3/2} \int d\mathbf{r} \, \phi(\mathbf{r}) \sin(\mathbf{p} \cdot \mathbf{r}) = -[\phi(\mathbf{r})]_{(I)}^T(-\mathbf{p})$$

which show that both the real and imaginary components of $\chi(\mathbf{p})$ are eigenfunctions of the inversion operator. Similarly, in the case when $\phi(\mathbf{r})$ is purely imaginary, one can write

$$\chi_{(R)}(\mathbf{p}) = (2\pi)^{-3/2} \int d\mathbf{r} \, \phi_{(I)}(\mathbf{r}) \sin(\mathbf{p} \cdot \mathbf{r}) = -[\phi(\mathbf{r})]_{(R)}^T(-\mathbf{p})$$

$$\chi_{(I)}(\mathbf{p}) = (2\pi)^{-3/2} \int d\mathbf{r} \, \phi_{(I)}(\mathbf{r}) \cos(\mathbf{p} \cdot \mathbf{r}) = [\phi(\mathbf{r})]_{(I)}^T(-\mathbf{p})$$

with the same conclusion.

The symmetry point groups to be used when working within the framework of the **p**-representation are easily obtained, either because they are identical to the corresponding groups in the **r**-representation (when these already contain the inversion operation) or by making use of the rules for the products of symmetry operations. Defranceschi and Berthier (1990) have summarized the correspondence between the position and momentum symmetry point groups for the most important cases.

Although the electron momentum density (*EMD*) distributions are discussed below, it is convenient to complete the present discussion with an observation at the orbital level. The EMD is given by

$$\pi(\mathbf{p}) = \chi^*(\mathbf{p})\chi(\mathbf{p}) = \chi^2_{(R)}(\mathbf{p}) + \chi^2_{(I)}(\mathbf{p})$$

so that, for example, when $\phi(\mathbf{r})$ is real, we obtain

$$\pi(\mathbf{p}) = ([\phi(\mathbf{r})]^T_{(R)}(-p))^2 + ([\phi(\mathbf{r})]^T_{(I)}(-p))^2 = \pi(-\mathbf{p})$$

which shows that the momentum density presents an inversion centre (Defranceschi and Berthier 1990). This fact ensures that the function $\phi(\mathbf{r})$, corresponding to $\chi(\mathbf{p})$, has no translational motion (Kaijser and Smith 1976).

4.4.2 Momentum Density Distributions

Electron momentum density (*EMD*) distributions are of particular interest because of their direct relationship with the results of experimental measurements. Electron-momentum spectroscopy (McCarthy and Weigold 1991) will provide an estimate of the probability of finding an electron in a given energy-momentum range, the spectrum of Compton scattered radiation (DuMond 1929, 1930, 1933 and Cooper 1985) will determine a particular projection of the electron momentum distribution of the system under study, and the results of angle-resolved photoelectron spectroscopy (Courths and Hüfner 1984) allows us to express the band dispersion relations in terms of the energy and crystal momentum for electrons in a single crystal. [See also the work of Duncanson and Coulson (1945), Epstein and Tanner (1977), Mendelsohn and Smith (1977), Leung and Brion (1985), Rawling and Davidson (1985), Brion (1986), and Vos and McCarthy (1997) as well as most of the references given below in this section.]

The momentum density, $\pi(\mathbf{p})$, for an *N*-electron system is defined (Simas *et al.* 1984a) as

$$\pi(\mathbf{p}) = N \int \psi^*(\gamma_1, \gamma_2, ..., \gamma_N)\psi(\gamma_1, \gamma_2, ..., \gamma_N)d\gamma_2\, d\gamma_3 ... d\gamma_N d\sigma_1$$

where $\gamma_i = (\mathbf{p}_i, \sigma_i)$ denotes a combined momentum and spin coordinate. Spherical averaging (over all space) of $\pi(\mathbf{p})$ leads to the spherically-averaged momentum density, $\bar{\pi}(p)$, and the radial momentum density, I(p), which are related by

$$\int_0^\infty dp \, p^2 \int_0^\pi d\theta_p \, \sin\theta_p \int_0^{2\pi} d\varphi_p \, \pi(\mathbf{p}) = \int_0^\infty dp \, \{4\pi p^2 \bar{\pi}(p)\} = \int_0^\infty dp \, I(p) = N$$

Within the impulse approximation (that is, for scattering by a single, independent electron, with neglect of the electron binding energy, and with a plane wave final electron state), the (intensity of the) Compton profile may be related to the *EMD* [DuMond (1929, 1930, 1933), Duncanson and Coulson (1945), and Kilby (1965); see also Benesch and Smith (1973), Epstein and Tanner (1977), Mendelsohn and Smith (1977), and Kulkarni *et al.* (1992)]. The directional Compton profile may be expressed as

$$J_{\mathbf{k}}(q) = \int d\mathbf{p} \, \pi(\mathbf{p})\delta(q - \mathbf{k}\cdot\mathbf{p}) \qquad \text{or} \qquad J(p_z) = \iint \pi(\mathbf{p}) dp_x dp_y$$

where $\mathbf{k}$ is the unit vector in the direction of the momentum transferred to the electron and q is the projection of the momentum-transfer vector on the initial electron-momentum vector. For isotropic systems as well as for those systems (liquids or gases) in which rotational averaging is essentially complete, the spherically-averaged Compton profile is given by

$$\bar{J}(q) = \frac{1}{2} \int_{|q|}^\infty dp \, p^{-1} I(p) \qquad\qquad\qquad\qquad (4.22)$$

in which case it is possible to obtain

$$I(p) = -2p\frac{d\bar{J}(q)}{dq}$$

from the Compton profile (Simas *et al.* 1984a).

Studies of *EMD* distributions have been performed for a variety of systems using the *FT* of functions determined in the **r**-representation within the framework of the independent-particle approximation at the *SCF* level.

In atoms, most of the work has focused on the monotomic *versus* non-monotonic behaviour of $\bar{\pi}(\mathbf{p})$ (Westgate *et al.* 1985), the topographical features of $\nabla^2\bar{\pi}(\mathbf{p})$ (Sagar *et al.* 1989), and the asymptotic behaviour of the *EMD* [see Bonham and Wellenstein (1977) and Thakkar *et al.* (1980) as well as the specific studies of Benesch and Smith (1973), Kaijser and Smith (1977), Mendelsohn and Smith (1977), Simas *et al.* (1982), Regier *et al.* (1985), Westgate *et al.* (1986), Thakkar (1987), and Thakkar *et al.* (1987)].

In molecules the attention has been centred particularly on the anisotropies of the *EMD* (and/or the related Compton profiles), starting with the pioneering work of Coulson [Coulson (1941ab), Coulson and Duncanson (1941), and Duncanson and Coulson (1941)], which led to the so-called *bond directional principle*, definitively established by Epstein and Tanner (1977) and reformulated by Tanner (1988). Specific contributions have been made by Henneker and Cade (1968),

Epstein (1970), Epstein and Lipscomb (1970), Smith 1971, Kaijser *et al.* (1973), Ahlenius and Lindner (1975), Kaijser and Lindner (1975), Tawil and Langhoff (1975), Langhoff and Tawil (1975), Kaijser and Smith (1976), Janis *et al.* (1978, 1979), Snyder and Weber (1978), Matcha and Pettit (1979), Kaijser *et al.* (1980), Koga and Morita (1981), Thakkar *et al.* (1981), Ramirez (1982), Sharma *et al.* (1983), Thakkar (1983), Trivedi and Steinborn 1983, Simas *et al.* (1984b), Thakkar (1984), Thakkar *et al.* (1984), Kulkarni *et al.* (1992), Kulkarni and Gadre (1993), and Wang *et al.* (1996).

Results obtained at the *SCF* level with a *3-21G* basis set will be used in order to illustrate graphically (with the respective atomic units used in all the figures) some of the comments made throughout this chapter.

Figures 4.1 and 4.2 present the electron densities of the degenerate e_{2g} *HOMO* of benzene (D_{6h} symmetry) in both the **r**- and **p**-representations, respectively. The two main observations made from these figures are that the **r**-density shows clearly the bond topology while the **p**-density is sharper. This last observation reflects the inverse relationship that exists between the two density representations: a highly localized **r**-density is associated with a broad **p**-density while a diffuse **r**-density transforms into a sharply peaked **p**-density. This contrast is, in fact, even more pronounced in the case of the a_{1g} *LUMO* of benzene, presented in Fig. 4.3.

A reverse situation exists for the core orbitals, for which the **r**-density is more concentrated than the **p**-density. A result of this situation is that the total **p**-density is far more diffuse than the total **r**-density, as observed in Figs. 4.4 and 4.5, which present the corresponding results for benzene and pyridine, respectively. As a consequence, the total **r**-densities show distinctly the symmetries of the two systems (D_{6h} for benzene and C_{2v} for pyridine) while one cannot discern the corresponding symmetries [D_{6h} for benzene and D_{2h} for pyridine (Defranceschi and Berthier 1990)] from the total **p**-densities.

4.4.3 Expectation Values

The functions in the **p**-representation are suitable for the calculation of the expectation values

$$< p^n > = \int d\mathbf{p} \, p^n \, \pi(\mathbf{p}) = 4\pi \int_0^\infty dp \, p^{n+2} \, \bar{\pi}(p) = \int_0^\infty dp \, p^n \, I(p) \tag{4.23}$$

some of which have physical significance [n = -1, isotropic Compton profile (see below); n = 1, shielding in nuclear magnetic resonance spectroscopy (Slichter 1963); n = 3, initial value of the Patterson function in *X*-ray crystallography (Glusker *et al.* 1986)] or are of theoretical interest (n = 0, test of the accuracy of the normalization of the function; n = 1, Slater-Dirac exchange energy in density functional theory (Parr and Yang 1989); n = 2, electron kinetic energy; n = 4, relativistic mass variation correction). These expectation values may be extracted from the experimental Compton profiles, according to

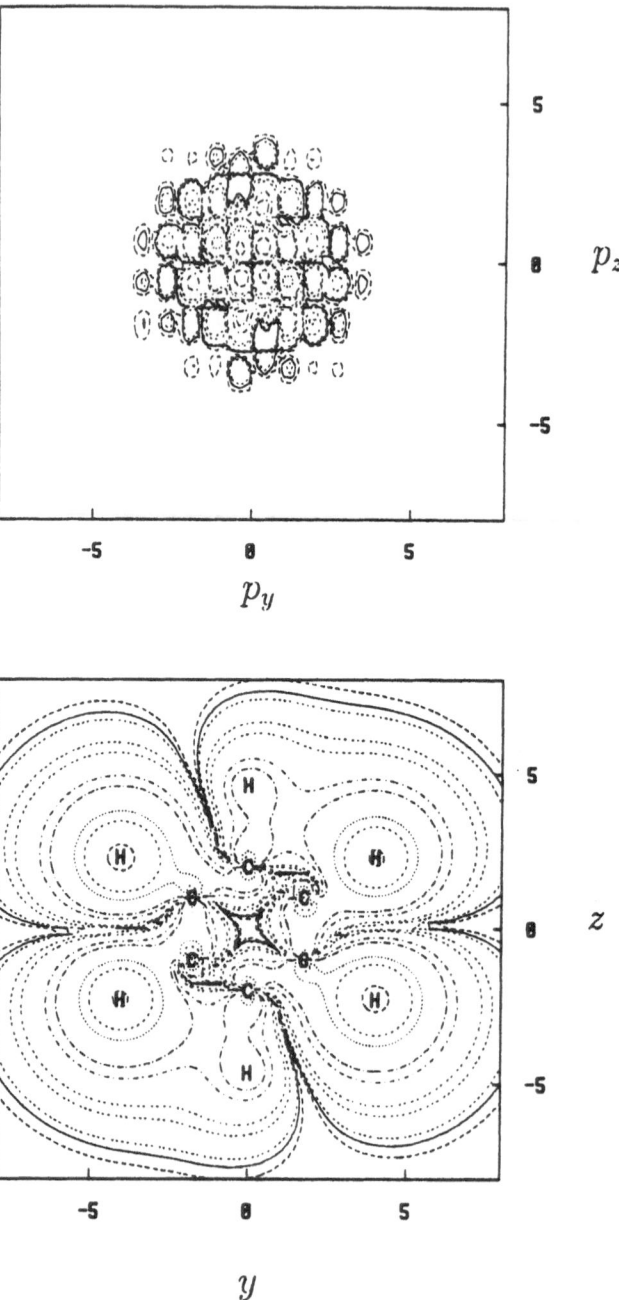

Figure 4.1. Two-dimensional electron momentum and position densities (in atomic units), on the molecular plane, for the $e_{2g}(x)$ *HOMO* of the benzene molecule.

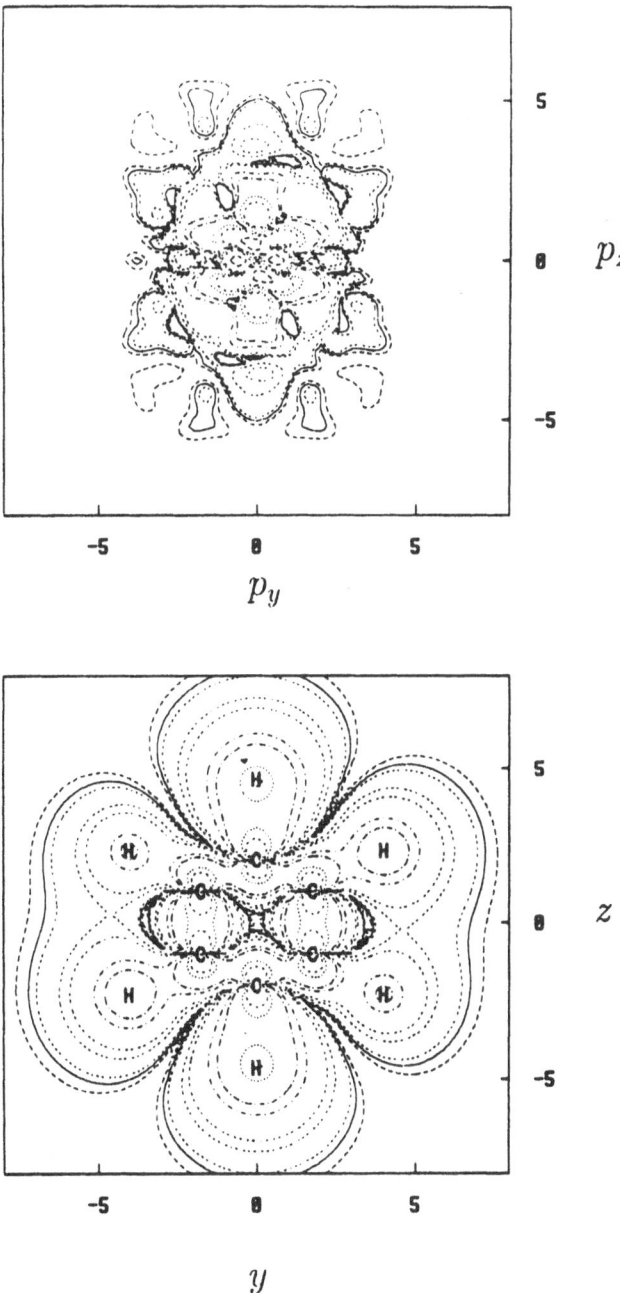

Figure 4.2. Two-dimensional electron momentum and position densities (in atomic units), on the molecular plane, for the $e_{2g}(y)$ *HOMO* of the benzene molecule.

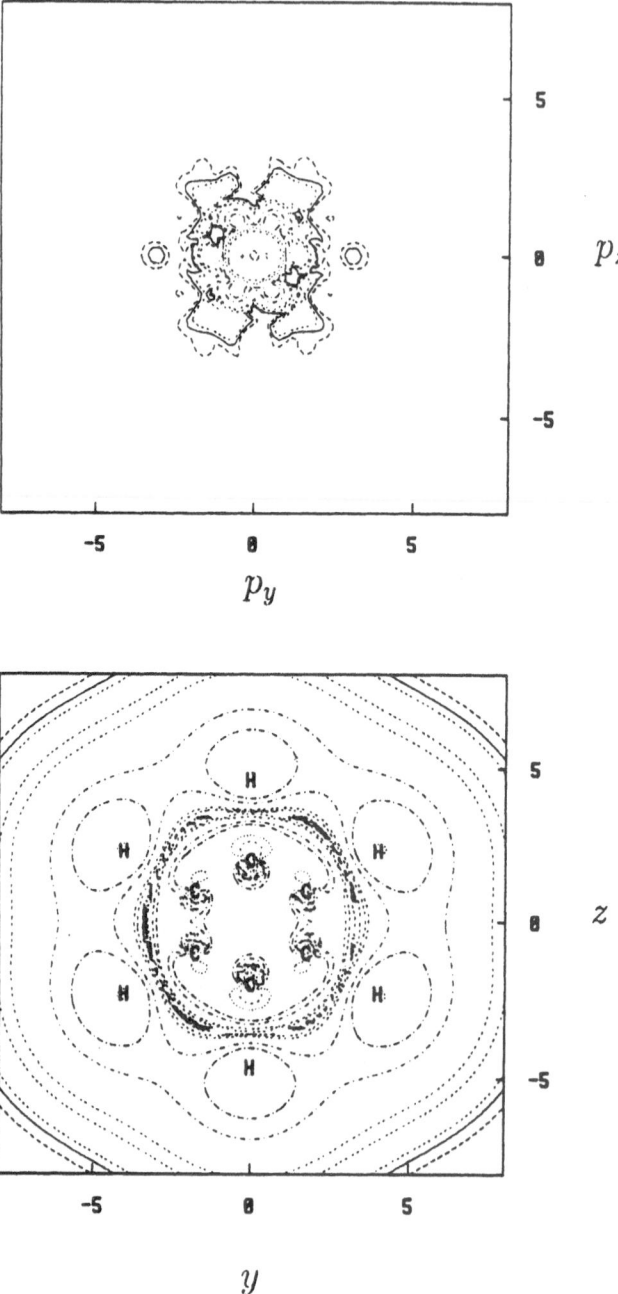

Figure 4.3. Two-dimensional electron momentum and position densities (in atomic units), on the molecular plane, for the a_{1g} *LUMO* of the benzene molecule.

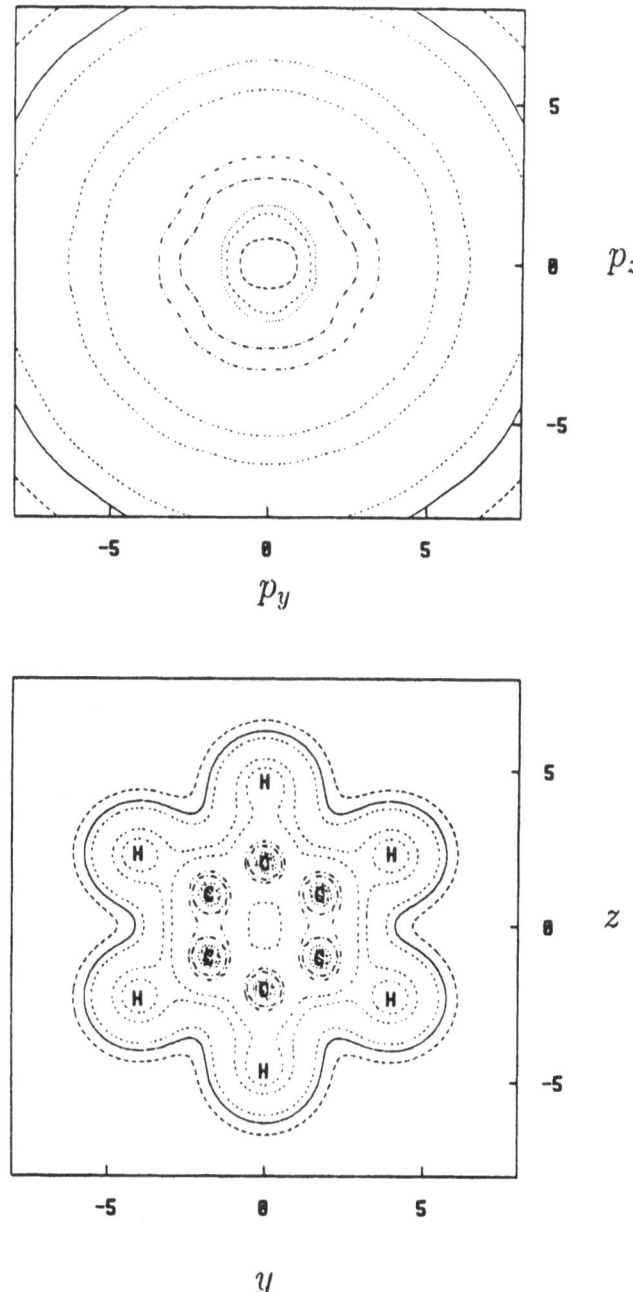

Figure 4.4. Total two-dimensional electron momentum and position densities (in atomic units), on the molecular plane, for the benzene molecule.

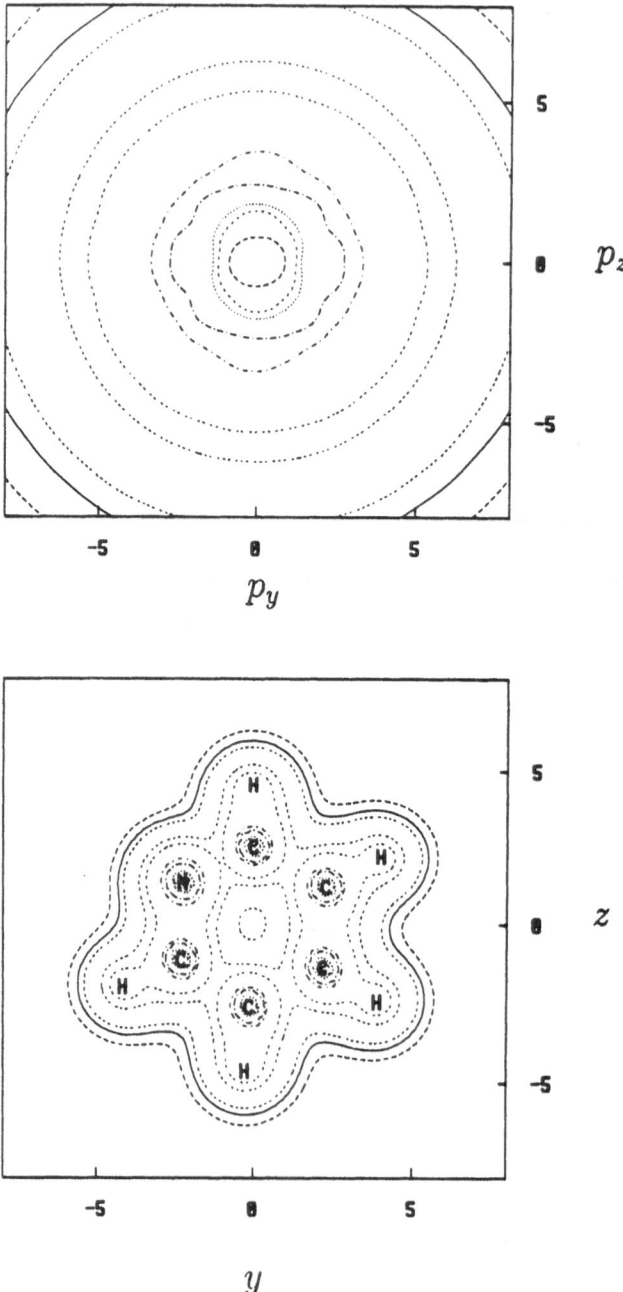

Figure 4.5. Total two-dimensional electron momentum and position densities (in atomic units), on the molecular plane, for the pyridine molecule.

$$< p^{-2} > = 2 \int_0^\infty dq \; q^{-2} \; [J(0) - J(q)]$$

$$< p^{-1} > = 2 \; [J(0)]$$

$$< p^n > = 2(n+1) \int_0^\infty dq \; q^n \; J(q)$$

(Epstein 1973, Thakkar *et al.* 1980, Gadre *et al.* 1983, Thakkar and Pedersen 1990).

Due to the interest of these values, systematic studies have been carried out both for atoms (Weiss *et al.* 1968, Biggs *et al.* 1975, Benesch 1975, Ponce 1983, Gadre *et al.* 1983, Westgate *et al.* 1991, Garcia de la Vega and Miguel 1993ab, 1994, 1995ab, and Koga and Thakkar 1996) and molecules (Pathak *et al.* 1986, Thakkar and Pedersen 1990, and Garcia de la Vega and Miguel 1997, 1998a).

An interesting application of the evaluation of expectation values $< p^n >$ concerns the test of the quality of basis sets (Thakkar *et al.* 1980, Simas *et al.* 1982, Regier *et al.* 1985, Garcia de la Vega and Miguel 1991, 1993c, 1995c). Such an application has a rather significant usefulness, as outlined below, and it is for this reason that a summary of results is presented here.

As repeatedly mentioned throughout the text, this work is concerned primarily with basic methodology. As a consequence, formulations based on perturbation theory and the variational principle as well as all those contributions, embodied in the so-called computational chemistry, have been ignored as a rule, the exception being the summary of reference on numerical approaches presented in Section 5.1. There is no doubt, however, that computational chemistry will develop even further, with practical goals in mind, and that basis sets of good quality will be needed in this endeavour. Therefore the availability of appropriate quality tests for existing as well as for still to be developed basis sets constitutes the first requirement for future work.

Usually, that quality is tested on the basis of energy results, either for atoms or for chosen molecules. Such an approach will provide an indication of the overall quality of the basis set (see Section 5.1) but not of its local quality in different regions of space. Such a knowledge is of fundamental importance because otherwise the risk exists of making poor predictions of given properties. That is, the quality of a basis set should be judged according to the results it yields not only for the energy but also for expectation values dependent on specific regions of space.

The expectation values $< r^n >$ and $< p^n >$ will provide that information and, consequently, it has been considered pertinent to offer here some illustrative results of the latter, taking into account the renewed interest in calculations in the **p**-representation.

Table 4.1. Expectation values $\langle p^n \rangle$ (in atomic units) for polyatomic molecules[a,b]

Molecule	p^{-2}	p^{-1}	p^1	p^2	p^3	p^4
			aug-cc-pVDZ basis set			
CH_4	.224571(2)	.100787(2)	.188285(2)	.802318(2)	.689442(3)	.113159(5)
$CH{\equiv}CH$	.250250(2)	.127861(2)	.315385(2)	.153216(3)	.137202(4)	.226940(5)
$CH_2{=}CH_2$	.335301(2)	.150727(2)	.336752(2)	.155718(3)	.137366(4)	.226597(5)
$CH_3{-}CH_3$	.359464(2)	.172289(2)	.358551(2)	.158274(3)	.137507(4)	.226192(5)
$CH_3C{\equiv}CH$	.537912(2)	.198379(2)	.485311(2)	.231135(3)	.205728(4)	.339975(5)
			cc-pVQZ basis set			
CH_4	.176559(2)	.100764(2)	.188474(2)	.803288(2)	.689850(3)	.115258(5)
$CH{\equiv}CH$	.202135(2)	.127571(2)	.315636(2)	.153235(3)	.137106(4)	.230865(5)
$CH_2{=}CH_2$	.246363(2)	.150393(2)	.337029(2)	.155784(3)	.137343(4)	.230667(5)
$CH_3{-}CH_3$	.294107(2)	.172232(2)	.358832(2)	.158401(3)	.137558(4)	.230396(5)
$CH_3C{\equiv}CH$	.319350(2)	.198138(2)	.485639(2)	.231168(3)	.205611(4)	.345922(5)
			aug-cc-pVQZ basis set			
CH_4	.157645(2)	.100759(2)	.188474(2)	.803283(2)	.689865(3)	.115266(5)
$CH{\equiv}CH$	.275687(2)	.127783(2)	.315593(2)	.153231(3)	.137109(4)	.230876(5)
$CH_2{=}CH_2$	.243206(2)	.150621(2)	.336979(2)	.155779(3)	.137349(4)	.230692(5)
$CH_3{-}CH_3$	.258108(2)	.172237(2)	.358832(2)	.158400(3)	.137561(4)	.230408(5)
$CH_3C{\equiv}CH$	.382727(2)	.198294(2)	.485601(2)	.231163(3)	.205615(4)	.345942(5)

[a]Obtained at the *SCF* level. See the text for details.
[b]The figures in parentheses denote powers of ten.

Table 4.2. Expectation values $<p^{-2}>$ (in atomic units) for some diatomic molecules[a,b]

Molecule	B3P86	B3PW91	BP86	BPW91
BO	.16098952(2)	.16132006(2)	.16380841(2)	.16331964(2)
BO⁻	.32607928(2)	.32674986(2)	.33943833(2)	.33594231(2)
BO⁺	.11270691(2)	.11147592(2)	.11392626(2)	.11226921(2)
CH	.14585348(2)	.14568579(2)	.14836408(2)	.14726367(2)
CH⁻	.20569680(2)	.20554068(2)	.21064977(2)	.20887219(2)
CH⁺	.10347161(2)	.10385126(2)	.10470163(2)	.10466266(2)
NO	.15875304(2)	.15847256(2)	.16053765(2)	.15970250(2)
NO⁻	.18993470(2)	.18936138(2)	.19318428(2)	.19164548(2)
NO⁺	.11706415(2)	.11681208(2)	.11784789(2)	.11740228(2)
OF	.13187553(2)	.13164680(2)	.13290710(2)	.13237228(2)
OF⁻	.15071129(2)	.15050915(2)	.15157387(2)	.15108160(2)
OF⁺	.11611274(2)	.11600573(2)	.11694690(2)	.11659039(2)
PH	.21405112(2)	.21456826(2)	.21553715(2)	.21508279(2)
PH⁻	.27611472(2)	.27608676(2)	.27810763(2)	.27639204(2)
PH⁺	.16520245(2)	.16581762(2)	.16593790(2)	.16602655(2)
SiH	.24705790(2)	.24791269(2)	.24873886(2)	.24829331(2)
SiH⁻	.32932872(2)	.32904271(2)	.33178036(2)	.32952921(2)
SiH⁺	.18728914(2)	.18806197(2)	.18809902(2)	.18814160(2)

[a]Obtained at the *DFT* level. See the text for details.
[b]The figures in parentheses denote powers of ten.

Table 4.3. Expectation values $<p^{-1}>$ (in atomic units) for some diatomic molecules[a,b]

Molecule	B3P86	B3PW91	BP86	BPW91
BO	.10013603(2)	.10011855(2)	.10061618(2)	.10038654(2)
BO⁻	.12725673(2)	.12721211(2)	.12809522(2)	.12772580(2)
BO⁺	.82435203(1)	.82446255(1)	.82662430(1)	.82564368(1)
CH	.68364492(1)	.68330375(1)	.68571501(1)	.68393384(1)
CH⁻	.85897270(1)	.85838288(1)	.86207514(1)	.85975057(1)
CH⁺	.53870044(1)	.53883522(1)	.53954349(1)	.53883589(1)
NO	.10648947(2)	.10644861(2)	.10675739(2)	.10657370(2)
NO⁻	.11942279(2)	.11935830(2)	.11985147(2)	.11962303(2)
NO⁺	.89836726(1)	.89822798(1)	.89963449(1)	.89857434(1)
OF	.10728959(2)	.10725122(2)	.10764602(2)	.10749842(2)
OF⁻	.11894574(2)	.11890065(2)	.11936104(2)	.11919664(2)
OF⁺	.97029509(1)	.97016607(1)	.97242298(1)	.97144423(1)
PH	.11343441(2)	.11338017(2)	.11364465(2)	.11337467(2)
PH⁻	.13375837(2)	.13372797(2)	.13406067(2)	.13378991(2)
PH⁺	.95719667(1)	.95726690(1)	.95813304(1)	.95683326(1)
SiH	.11314568(2)	.11311624(2)	.11339709(2)	.11310505(2)
SiH⁻	.13644485(2)	.13639164(2)	.13686504(2)	.13648331(2)
SiH⁺	.93989722(1)	.94027209(1)	.94087039(1)	.93973786(1)

[a]Obtained at the *DFT* level. See the text for details.
[b]The figures in parentheses denote powers of ten.

Table 4.4. Expectation values <p^1> (in atomic units) for some diatomic molecules[a,b]

Molecule	B3P86	B3PW91	BP86	BPW91
BO	.35068894(2)	.35067752(2)	.35057531(2)	.35073638(2)
BO$^-$	.35574838(2)	.35571888(2)	.35553884(2)	.35570216(2)
BO$^+$	.34199468(2)	.34196536(2)	.34203059(2)	.34213250(2)
CH	.15537023(2)	.15536374(2)	.15528539(2)	.15538861(2)
CH$^-$	.16195547(2)	.16195085(2)	.16182922(2)	.16194599(2)
CH$^+$	.14633382(2)	.14631066(2)	.14631324(2)	.14636492(2)
NO	.42982451(2)	.42981628(2)	.42988595(2)	.43005086(2)
NO$^-$	.43986650(2)	.43987042(2)	.43973087(2)	.43992354(2)
NO$^+$	.42345855(2)	.42344659(2)	.42355179(2)	.42367601(2)
OF	.53119476(2)	.53120900(2)	.53102728(2)	.53122703(2)
OF$^-$	.54132294(2)	.54133586(2)	.54109753(2)	.54130900(2)
OF$^+$	.51879532(2)	.51879225(2)	.51884902(2)	.51899975(2)
PH	.67080001(2)	.67080977(2)	.67039726(2)	.67056235(2)
PH$^-$	.67632232(2)	.67630839(2)	.67587918(2)	.67602823(2)
PH$^+$	.66382675(2)	.66382523(2)	.66348944(2)	.66360969(2)
SiH	.60217406(2)	.60217521(2)	.60175986(2)	.60191349(2)
SiH$^-$	.60700368(2)	.60699638(2)	.60652803(2)	.60668931(2)
SiH$^+$	.59540334(2)	.59540001(2)	.59505840(2)	.59516795(2)

[a]Obtained at the *DFT* level. See the text for details.
[b]The figures in parentheses denote powers of ten.

Table 4.5. Expectation values $<p^2>$ (in atomic units) for some diatomic molecules[a,b]

Molecule	B3P86	B3PW91	BP86	BPW91
BO	.19927029(3)	.19920464(3)	.19945472(3)	.19943141(3)
BO$^-$	.19969347(3)	.19962077(3)	.19985190(3)	.19982667(3)
BO$^+$	.19821959(3)	.19815229(3)	.19846912(3)	.19843690(3)
CH	.76612132(2)	.76582791(2)	.76659314(2)	.76658561(2)
CH$^-$	.76867519(2)	.76834230(2)	.76922591(2)	.76916941(2)
CH$^+$	.75856062(2)	.75825257(2)	.75909529(2)	.75901568(2)
NO	.25884046(3)	.25875279(3)	.25910687(3)	.25906885(3)
NO$^-$	.25978730(3)	.25969935(3)	.26000337(3)	.25996613(3)
NO$^+$	.25867507(3)	.25858610(3)	.25895614(3)	.25890661(3)
OF	.34882904(3)	.34873155(3)	.34909245(3)	.34905122(3)
OF$^-$	.34962244(3)	.34952064(3)	.34988543(3)	.34984128(3)
OF$^+$	.34738772(3)	.34728968(3)	.34771800(3)	.34766951(3)
PH	.68148762(3)	.68147938(3)	.68114531(3)	.68119963(3)
PH$^-$	.68179863(3)	.68177830(3)	.68145864(3)	.68149886(3)
PH$^+$	.68085800(3)	.68085557(3)	.68052858(3)	.68058483(3)
SiH	.57800516(3)	.57800108(3)	.57769076(3)	.57774597(3)
SiH$^-$	.57820415(3)	.57819069(3)	.57788820(3)	.57793285(3)
SiH$^+$	.57738782(3)	.57738981(3)	.57708679(3)	.57714318(3)

[a]Obtained at the *DFT* level. See the text for details.
[b]The figures in parentheses denote powers of ten.

Table 4.6. Expectation values $<p^3>$ (in atomic units) for some diatomic molecules[a,b]

Molecule	B3P86	B3PW91	BP86	BPW91
BO	.21217241(4)	.21203706(4)	.21256305(4)	.21242123(4)
BO $^-$	.21222720(4)	.21209024(4)	.21260153(4)	.21246196(4)
BO$^+$	.21217634(4)	.21204278(4)	.21260172(4)	.21246143(4)
CH	.69203473(3)	.69157785(3)	.69324799(3)	.69285288(3)
CH $^-$	.68909323(3)	.68860738(3)	.69043417(3)	.68998300(3)
CH$^+$	.69513238(3)	.69466312(3)	.69631571(3)	.69590815(3)
NO	.28768754(4)	.28749746(4)	.28817784(4)	.28798161(4)
NO $^-$	.28718295(4)	.28699084(4)	.28767241(4)	.28747120(4)
NO$^+$	.28775288(4)	.28755655(4)	.28827084(4)	.28806461(4)
OF	.42977378(4)	.42950368(4)	.43044165(4)	.43015133(4)
OF $^-$	.42908389(4)	.42881076(4)	.42976602(4)	.42946956(4)
OF$^+$	.43050088(4)	.43023163(4)	.43117725(4)	.43088996(4)
PH	.13452924(5)	.13452557(5)	.13449852(5)	.13449922(5)
PH $^-$	.13450114(5)	.13449623(5)	.13447215(5)	.13447081(5)
PH$^+$	.13456364(5)	.13456150(5)	.13453286(5)	.13453569(5)
SiH	.10747461(5)	.10747213(5)	.10744784(5)	.10744950(5)
SiH $^-$	.10744354(5)	.10743973(5)	.10741865(5)	.10741823(5)
SiH$^+$	.10750704(5)	.10750587(5)	.10748052(5)	.10748405(5)

[a]Obtained at the *DFT* level. See the text for details.
[b]The figures in parentheses denote powers of ten.

Table 4.7. Expectation values $<p^4>$ (in atomic units) for some diatomic molecules[a,b]

Molecule	B3P86	B3PW91	BP86	BPW91
BO	.43911128(5)	.43871798(5)	.44022248(5)	.43976143(5)
BO⁻	.43901449(5)	.43862224(5)	.44010314(5)	.43964857(5)
BO⁺	.43966784(5)	.43927885(5)	.44080047(5)	.44034673(5)
CH	.11680399(5)	.11670327(5)	.11712230(5)	.11701737(5)
CH⁻	.11622961(5)	.11612680(5)	.11656469(5)	.11645352(5)
CH⁺	.11762835(5)	.11752125(5)	.11794165(5)	.11783111(5)
NO	.60701843(5)	.60646091(5)	.60844731(5)	.60780656(5)
NO⁻	.60536718(5)	.60480721(5)	.60682584(5)	.60617568(5)
NO⁺	.60702580(5)	.60645195(5)	.60852244(5)	.60786252(5)
OF	.10148178(6)	.10138924(6)	.10170240(6)	.10159318(6)
OF⁻	.10126828(6)	.10117550(6)	.10149136(6)	.10138099(6)
OF⁺	.10176279(6)	.10166987(6)	.10198024(6)	.10187161(6)
PH	.52986037(6)	.52984320(6)	.52988309(6)	.52987179(6)
PH⁻	.52968935(6)	.52966919(6)	.52971820(6)	.52970062(6)
PH⁺	.53012867(6)	.53011818(6)	.53014987(6)	.53014868(6)
SiH	.39754025(6)	.39752911(6)	.39753877(6)	.39753381(6)
SiH⁻	.39738348(6)	.39736780(6)	.39738946(6)	.39737696(6)
SiH⁺	.39777396(6)	.39776846(6)	.39777264(6)	.39777681(6)

[a]Obtained at the *DFT* level. See the text for details.
[b]The figures in parentheses denote powers of ten.

All the results have been obtained, using existing software (Garcia de la Vega and Miguel 1998b), for groundstate functions of both diatomics and polyatomics (at the experimental equilibrium geometry), determined within the framework of either the *SCF* or the *DFT* formalisms.

The *SCF* results for polyatomics, presented in Table 4.1, were obtained with the basis sets *cc*-p*VQZ*, *aug-cc*-p*VDZ*, and *aug-cc*-p*VQZ* (Dunning 1989, Kendall *et al.* 1992, Woon and Dunning 1993). They confirm that the values for $n > 0$ increase as the basis set is improved while the values for $n < 0$ decrease, as already observed for diatomics (Garcia de la Vega and Miguel 1998a).

On the other hand, the *DFT* results for some chosen diatomics, as well as for their anions and cations, were obtained using exclusively the basis set *cc*-p*VQZ*, with the functionals *BP86, BPW91, B3P86*, and *B3PW91* (Perdew 1986ab, Perdew and Wang 1992, Becke 1992ab; see also Parr and Yang 1989, 1995). The main observation made from these results, presented in Tables 4.2-4.7, is that the magnitudes of the expectation values, for any of the functionals tested, vary in the order anion > neutral molecule > cation, for $n = -2, -1, 1, 2$, and in the order cation > neutral molecule > anion, for $n = 3, 4$. This behaviour is related to the values of the contributions of the individual orbitals as well as to the charge of the system. On one hand, the magnitudes of the orbital contributions vary, as a rule, in the order inner orbitals > valence orbitals, for $n > 0$, and valence orbitals > inner orbitals, for $n < 0$ (see, e.g., the work of Garcia de la Vega and Miguel 1998a). The effect of the charge of the system is due to the greater occupation of the valence orbitals in the anion and to the smaller size of the electron cloud in the cations. There is also a dependence of the magnitudes of the expectation values on the functional used but a larger set of results may be needed in order to make a conclusive observation.

5 The Local Schrödinger Equation

5.1 Alternatives for the Computational Solution of the Schrödinger Equation

The analytical solution of the Schrödinger equation, as already mentioned in Section 4.3, is only possible in the case of one-electron systems and recourse must be made of approximations when dealing with many-electron systems.

Most often the calculations are based on the variational principle, within the framework of the independent-particle model and using the expansion approximation. The functions obtained in such calculations will be denoted here as *computational functions*.

The treatments for the determination of computational functions lie outside of the scope of this work and, furthermore, they have been discussed extensively in the literature. Some comments are appropriate, however, in order to bring into perspective the reasons for the development of the alternative presented in this chapter: the local solution of the Schrödinger equation within the context of the so-called *local energy* (*LE*) methods. For completeness, a survey of the alternate numerical methods for global solutions will also be presented.

5.1.1 Computational Solutions of the Global Problem

The variational calculations for ground states are based on the criterion that the value

$$E = \frac{<\phi|\mathcal{H}|\phi>}{<\phi|\phi>} \tag{5.1}$$

corresponding to a trial function ϕ represents an upper bound to the true eigenvalue. The goal is to obtain, through manipulation of ϕ, a global value of E as low as possible, which will tend to the exact eigenvalue as the accuracy of the calculation is improved.

The correctness of the computational function depends on the computing capabilities available. Typically, orbitals are described by means of expansions in terms of basis functions (which, when developed in particular for the construction of molecular orbitals, are usually optimized for the isolated atoms). The predictive calculations have been rather successful in a number of cases but highly accurate calculations require large basis sets and, correspondingly, considerable computer time. In addition, many physical properties as well as diverse systems (such as, e.g., anions) are known to be highly sensitive to the quality of the basis sets.

An alternate, accurate approach (see, e.g., Defranceschi and Delhalle 1989) consists of the direct, numerical determination of the orbitals. This approach has proved to be extremely successful for atoms and diatomics but its implementation for polyatomics, in the r-representation, is faced with the problem of the Coulomb singularities. Such a problem may be eliminated by operating in the p-representation in which case, at least in principle, highly-accurate, multicentre calculations may be possible.

The characteristics and possible difficulties of the most important approaches, in both the r- and the p-representation, may be summarized as follows.

(a) Computational solutions in the r-representation.

For atoms, mono- and multi-configurational Hartree-Fock (Hartree 1957, Froese-Fischer 1977) as well as Dirac and Dirac-Fock (Grant 1970, Desclaux 1975) functions may be obtained almost routinely by numerical procedures using a one-dimensional grid with a few hundred points.

The first attempts at a seminumerical solution of the mono- and multi-configurational Hartree-Fock equations for diatomics were made by McCullough and Adamowicz. In their approach the orbitals were expressed by sums of products of associated Legendre functions in one dimension, with the second dimension represented by a numerical expansion and the third dimension treated analytically (McCullough 1974, 1975, 1981, 1986, Christiansen and McCullough 1977, Adamowicz and McCullough 1981, Adamowicz and Bartlett 1986, 1988, Adamowicz 1988, Richman and McCullough 1988). Other treatments have involved the use of splines for the various local density functionals (Becke 1982, 1983, 1985, 1986, 1988), a finite-element method for the solution of Poisson, Schrödinger, Hartree-Fock, Hartree-Fock-Slater, and Dirac-Slater equations (Schulze and Kolb 1985, Heinemann *et al.* 1987, 1988*ab*, Sundholm 1988, Yang *et al.* 1992, 1993), and a finite-difference approach, in which the potentials are expressed in a two-dimensional, elliptical coordinate system and the solutions are obtained by a two-dimensional, numerical approach, with the third dimension treated analytically, and which has been applied to the solution of Hartree-Fock and Hartree-Fock-Slater equations (Laaksonen *et al.* 1986, 1988, Pyykkö *et al.* 1987, Müller-Plathe and Laaksonen 1989, Kobus 1993, 1994, Kobus *et al.* 1994, 1995, Moncrieff *et al.* 1995, Kobus *et al.* 1996) as well as to the solution of Dirac and Dirac-Slater equations (Laaksonen and Grant 1984*ab*, Sundholm 1988).

A three-dimensional approach for polyatomics (Becke 1988), using local, one-centre expansions at every nucleus, has been applied to the solution of both the Poisson (Becke and Dickson 1988) and the Schrödinger (Becke and Dickinson 1990) equations, using an iterative matrix method. Molecular properties may be obtained using a basis-set-free algorithm (Dickinson and Becke 1993), which could play an important role in the future.

(b) Computational solutions in the **p**-representation.

An important characteristic of the **p**-representation is the fact that the Coulomb singularities appearing in the equations in the **r**-representation are transferred to infinity. The transformed equations have a single singularity at the origin, which suggests the use of momentum-space polar coordinates. Another significant observation is that the largest amplitudes of the functions are centred at the origin of the momentum space.

Significant contributions in connection with the determination of functions in the **p**-representation have been the treatment of the discrete and continuous spectra of one-electron atoms (see Bethe and Salpeter 1957) and its extension to many-electron systems (McWeeny 1949 and McWeeny and Coulson 1949), the expansion of the functions as linear combinations of four-dimensional spherical harmonics (Shibuya and Wulfman 1965), the solution of the Fourier Transform of the Hartree-Fock equations using an appropriate integration spherical grid (Lasettre 1973, Navaza and Tsoucaris 1981), the examination of the behaviour and the properties of the electronic functions (Lasettre 1989), the revision of the self-consistent field equations for atoms and polymer chains (Berthier *et al.* 1990), the iterative procedure of Avery and co-workers (Avery and Wen 1982, Avery *et al.* 1986, Avery 1987, Avery and Larsen 1988, Avery 1989, Avery and Antonsen 1992, Avery and Herschbach 1992, Avery and Hansen 1996, Avery *et al.* 1996, and Aquilanti and Avery 1997), and the study of linear H_n chains and other complex systems (Defranceschi *et al.* 1984*ab*, Defranceschi and Delhalle 1986, Delhalle and Defranceschi 1987, Berthier *et al.* 1985, 1989, 1997, Calais *et al.* 1992, Defranceschi *et al.* 1992, DeWindt *et al.* 1993*ab*).

The first attempt to solve numerically the momentum-space equations was due to Fock (1935), by means of a hypergeometrical transformation of the momentum space into a hypersphere. This transformation was later applied to the Hartree-Fock equations (Shibuya and Wulfman 1965, Novosadov 1976, Monkhorst and Jeziorski 1979, Novosadov and Pogonin 1982, Rodriguez and Ishikawa 1988*ab*) and to the development of perturbative methods for the solution of the problem of a single electron moving in the Coulombic field of two nuclei (Koga 1985*ab*, Koga and Kawa-ai 1986, Koga and Matuhashi 1987, 1988, Koga *et al.* 1988, 1991, Koga and Ougihara 1989) and of the van der Waals interaction of two Hydrogen atoms (Koga and Matsumoto 1985, Koga and Ujiie 1986, Koga *et al.* 1987, Koga and Yamazaki 1989). The Fast Fourier Transform approach has also been used for full numerical treatments (Navaza and Tsoucaris 1981, Alexander and Monkhorst 1987, Alexander *et al.* 1988). Recently, Fischer and Defranceschi (1994, 1998) have used orthonormal wavelets to decompose the Schrödinger operator in a non-standard form and an iterative procedure to solve the Hydrogen atom.

5.1.2 Local vs. Global Computational Solutions

The variational calculations, based on Eq. (5.1), optimize the trial function ϕ in a restricted region of the configuration space, directly connected with the Hamiltonian operator. The resulting function, obtained by such a procedure, does not lead necessarily to equally satisfactory values for the expectation values of other observables (Pilar 1968), corresponding to operators which may emphasize different regions of that space.

Because of its global character, the solution does not provide any information on the local inadequacies. The alternative, as first proposed by Bartlett and Frost, consists of determining the value of the function at every point in the electron configuration space *(PECS)*, on the basis that for the exact function the ratio $\mathcal{H}\Psi/\Psi$ must be the same for every *PECS*. For an approximate function ϕ, the ratio $\mathcal{H}\phi/\phi$ will not be constant for every *PECS* and the local energy (*LE*) methods will try to obtain the best function that minimizes the variance in the local energy computed on a finite sample grid (with arbitrary weighting of the grid points). Thus, the evaluation of integrals in the global calculations is replaced with the evaluation of weighted sums over the grid.

The sample grid must meet some minimal requirements. The number, distribution, and weighting of the sample points must be such that, when operating within the context of the expansion approximation, the basis functions will form a linearly independent set over the sample grid as well as over the whole coordinate space.

5.2 The Local Energy Methods

The *LE* approach was suggested by Weinstein (1934), Bartlett (1937, 1955), and Frost (1942) and subsequently developed by Frost and co-workers (Frost *et al.* 1960, 1961, Frost 1962, Gimarc and Frost 1963, Harriss and Frost 1964), with additional contributions from Conroy (Conroy 1964, Conroy and Bruner 1965), Stanton and Taylor (1966), and Rourke and Stewart (1967, 1968). Variations and extensions of the original approach were then proposed by Rothstein, Grelland, and Palke.

5.2.1 The Frost Equation

Let us consider a trial function $\phi(\mathbf{r},\sigma)$, where $\mathbf{r}$ and σ denote the spatial and spin variables, $\{\mathbf{r}_1,\mathbf{r}_2,...,\mathbf{r}_N\}$ and $\{\sigma_1,\sigma_2,...,\sigma_N\}$, of the electrons in the system, respectively. The local energy at the p-th PECS, characterized by r_p [i.e., with $r_1(p),r_2(p),...,r_N(p)$], is defined as

$$\varepsilon_p \equiv \varepsilon(r_p) = \frac{\mathcal{H}\phi(r_p,\sigma)}{\phi(r_p,\sigma)} \tag{5.2}$$

where $\phi(r_p,\sigma)$ and $\mathcal{H}\phi(r_p,\sigma)$ are obtained by using in $\phi(r,\sigma)$ and $\mathcal{H}\phi(r,\sigma)$, respectively, the values of the spatial coordinates of the electrons at the *PECS* under consideration. Hereafter, for simplicity, it will be assumed that $\phi(r,\sigma)$ is real and that the Hamiltonian operator is independent of the spin.

When considering a set of *PECS*, the variance of the local energy is given (Frost 1942) by

$$V = <\varepsilon^2> - <\varepsilon>^2 \tag{5.3}$$

where the bra-ket notation is used to denote the mean values

$$<\varepsilon> = \frac{\sum_p g_p \varepsilon_p}{\sum_p g_p} \tag{5.4a}$$

$$<\varepsilon^2> = \frac{\sum_p g_p \varepsilon_p^2}{\sum_p g_p} \tag{5.4b}$$

The weighting factors g_p may be taken

$$g_p = w_p \phi^2(r_p,\sigma) \tag{5.5}$$

to be proportional to the amplitude of the function at each *PECS*, with the factor w_p related to the volume element (Frost *et al.* 1960). Therefore one can write

$$<\varepsilon> = \frac{\sum_p w_p \phi^2(r_p,\sigma)[\mathcal{H}\phi(r_p,\sigma)/\phi(r_p,\sigma)]}{\sum_p w_p \phi^2(r_p,\sigma)} = \frac{\sum_p w_p \phi(r_p,\sigma)\mathcal{H}\phi(r_p,\sigma)}{\sum_p w_p \phi^2(r_p,\sigma)} \tag{5.6a}$$

$$<\varepsilon^2> = \frac{\sum_p w_p \phi^2(r_p,\sigma)[\mathcal{H}\phi(r_p,\sigma)/\phi(r_p,\sigma)]^2}{\sum_p w_p \phi^2(r_p,\sigma)} = \frac{\sum_p w_p [\mathcal{H}(r_p,\sigma)]^2}{\sum_p w_p \phi^2(r_p,\sigma)} \tag{5.6b}$$

It can be seen that the expression for $<\varepsilon>$, Eq. (5.6a), is analogous to the expression, Eq. (5.1), used in the variational calculations and this analogy may be used in order to take care of the spin. If $\phi(\mathbf{r},\sigma)$ is given (as in the case of one-electron systems, etc.) as the product of a spatial component and a spin component, then the latter may be omitted in Eqs. (5.6). If, however, $\phi(\mathbf{r},\sigma)$ is given as a sum of products of spatial and spin functions, a cancellation of terms is made as in the variational expression (Rourke and Stewart 1967). Therefore one can write, in general,

$$<\varepsilon> = \frac{\sum\limits_{p} w_p \phi(\mathbf{r}_p) \mathcal{H}\phi(\mathbf{r}_p)}{\sum\limits_{p} w_p \phi^2(\mathbf{r}_p)} \tag{5.7a}$$

$$<\varepsilon^2> = \frac{\sum\limits_{p} w_p [\mathcal{H}\phi(\mathbf{r}_p)]^2}{\sum\limits_{p} w_p \phi^2(\mathbf{r}_p)} \tag{5.7b}$$

with the understanding that $\phi(\mathbf{r}_p)\mathcal{H}\phi(\mathbf{r}_p)$, $\mathcal{H}\phi(\mathbf{r}_p)$, and $\phi^2(\mathbf{r}_p)$ may consist of more than one term.

The trial function $\phi(\mathbf{r})$ may be expanded

$$\phi(\mathbf{r}) = \sum_{i} \psi_i(\mathbf{r}) c_i$$

in terms of known functions $\psi_i(\mathbf{r})$ and coefficients c_i to be determined (Frost 1942). One can then write (Stanton and Taylor 1966)

$$\sum_{p} w_p \phi^2(\mathbf{r}_p) = \sum_{p} w_p \{ \sum_{i} \sum_{j} c_i \psi_i(\mathbf{r}_p)\psi_j(\mathbf{r}_p) c_j \}$$

$$= \sum_{i} \sum_{j} c_i \{ \sum_{p} w_p \psi_i(\mathbf{r}_p)\psi_j(\mathbf{r}_p) \} c_j = \sum_{i} \sum_{j} c_i S_{ij} c_j = \mathbf{c}^\dagger \mathbf{S} \mathbf{c}$$

$$\sum_{p} w_p \phi(\mathbf{r}_p)\mathcal{H}\phi(\mathbf{r}_p) = \frac{1}{2} \sum_{p} w_p \{ \sum_{i} \sum_{j} [c_i \psi_i(\mathbf{r}_p)\mathcal{H}\psi_j(\mathbf{r}_p) c_j + c_j \psi_j(\mathbf{r}_p)\mathcal{H}\psi_i(\mathbf{r}_p) c_i \}$$

$$= \frac{1}{2} \sum_{i} \sum_{j} c_i \{ \sum_{p} [w_p \psi_i(\mathbf{r}_p)\mathcal{H}\psi_j(\mathbf{r}_p) + w_p \mathcal{H}\psi_i(\mathbf{r}_p)\psi_j(\mathbf{r}_p)] \} c_j$$

$$= \sum_{i} \sum_{j} c_i (H'_{ij} + H'_{ji}) c_j = \mathbf{c}^\dagger (\mathbf{H}' + \mathbf{H}'^\dagger) \mathbf{c} = \mathbf{c}^\dagger \mathbf{H} \mathbf{c}$$

$$\sum_p w_p[\mathcal{H}\phi(r_p)]^2 = \sum_p w_p\{\sum_i \sum_j c_i\mathcal{H}\psi_i(r_p)\mathcal{H}\psi_j(r_p)c_j\}$$

$$= \sum_i \sum_j c_i\{\sum_p w_p\mathcal{H}\psi_i(r_p)\mathcal{H}\psi_j(r_p)\}c_j = \sum_i \sum_j c_iG_{ij}c_j = c^\dagger Gc$$

where c is a column vector, with elements c_i, and S, H, and G are symmetric matrices, with elements (at the p-th *PECS*)

$$S_{ij} = \sum_p w_p\psi_i(r_p)\psi_j(r_p)$$

$$H_{ij} = \frac{1}{2}\sum_p w_p[\psi_i(r_p)\mathcal{H}\psi_j(r_p) + \mathcal{H}\psi_i(r_p)\psi_j(r_p)]$$

$$G_{ij} = \sum_p w_p\mathcal{H}\psi_i(r_p)\mathcal{H}\psi_j(r_p)$$

Introducing (Stanton and Taylor 1966) the normalization condition

$$c^\dagger Sc = 1$$

one can write

$$<\varepsilon> = c^\dagger Hc \qquad\qquad (5.8a)$$

$$<\varepsilon^2> = c^\dagger Gc \qquad\qquad (5.8b)$$

and [see Eq. (5.3)]

$$V = c^\dagger Gc - (c^\dagger Hc)^2 \qquad\qquad (5.8c)$$

which may minimized (Frost 1942) with respect to the expansion coefficients, subject to the above normalization condition. The resulting equation, obtained by standard variational techniques (using Lagrangian multipliers),

$$(G - 2<\varepsilon>H)c = \lambda Sc \qquad\qquad (5.9)$$

is the so-called Frost equation, which may be used for the determination of the optimum expansion coefficients (corresponding to minimum variance) and the subsequent evaluation of the local energy, according to Eq. (5.8a). Multiplication of both sides of this equation on the left by $c^\dagger$ yields, after reordering,

$$\lambda = V - <\varepsilon>^2 \qquad\qquad (5.10)$$

which shows the dependence of the eigenvalue λ on $<\varepsilon>$, which has implications in conjunction with the practical solution of the problem (see below).

A similar set of simultaneous equations is obtained for non-linear parameters (Harriss and Roubal 1968) and variations and improvements of the method are found in the local moments method (Silverstone *et al.* 1967), which combines the moments method (Vorobyev 1962) with the *LE* method, the orthonormalized *LE* method (Scott 1974), the linear non-symmetric eigenvalue equations (Carlson 1967, Harriss and Carlson 1969), which establishes the relationships with other numerical approaches (Frost *et al.* 1960, Goodisman 1965ab, Delos and Blinder 1967, Silverstone *et al.* 1967), and the evaluation of molecular properties (Davies 1968).

The critical steps in a *LE* calculation are the selection of the sample grid and the solution of the eigenvalue equations, while the excessive amount of computing time required constitutes a practical obstacle.

The selection of the sample grid is of importance insofar as it will affect the convergence of the procedure. A clear example of an inadequate selection is the one in which all the chosen *PECS* would have the same ε_p, in which case it would be $V = 0$ independently of the correctness of the trial function. Ehrenson and Harp (1973) use the so-called *importance sampling* procedure while Frost *et al.* (1960) chose the points in a systematic manner, using the method of Gaussian quadratures (Kopal 1955), and obtained local energies approaching the integrated value. Such an approach is satisfactory for systems with a reduced number of electrons but it would be expensive in all other cases because of the large number of points required in order to cover all the variables systematically. This difficulty suggests the use of a random procedure for the selection of the points. However, such a procedure will cause an uncertainty in the results, which will be reflected in a random fluctuation in the local energies obtained in repeated calculations. These random fluctuations might be removed by extrapolation to zero variance (Conroy 1964a) but contradictory experiences were reported by Michels (1966) and Rourke and Stewart (1967).

The usual techniques (involving both orthonormalization and diagonalization) for the solution of an eigenvalue problem would imply the use of the matrix equation

$$(G - 2 <\varepsilon> H)\,C = \lambda SC \qquad (5.11)$$

where C is now a matrix whose columns are the vectors c. When using n expansion functions, the procedure would yield n expansion vectors and n eigenvalues. The problem arises from the fact that the value of $<\varepsilon>$ only depends on a single expansion vector. This difficulty may be overcome either by an approximate method of solution of the set of simultaneous equations implicit in Eq. (5.9) (Frost *et al.* 1960) or by perturbation theory (Stanton and Taylor 1966).

5.2.2 Reduced Local Energies

The Frost local energy, defined in the preceding section, depends on the spatial variables of all the electrons in the system. The practical difficulties that such a dependence entails for systems with a large number of electrons may be avoided, although at a cost, through the alternate proposal of Rothstein and co-workers (Thomas *et al.* 1976, Javor *et al.* 1977) of a *reduced local energy (RLE)*, which will depend only on the spatial variables of a single electron.

The reduced local energy for an N-electron system is defined as

$$\varepsilon_R \equiv \varepsilon_R(\mathbf{r}_1) = \frac{\eta(\mathbf{r}_1)}{\rho(\mathbf{r}_1)} \tag{5.12}$$

where (with $d\tau_i = d\mathbf{r}_i d\sigma_i$)

$$\eta(\mathbf{r}_1) = \int \phi^*(\mathbf{r}_1,\mathbf{r}_2,...,\mathbf{r}_N)\mathcal{H}\phi(\mathbf{r}_1,\mathbf{r}_2,...,\mathbf{r}_N)d\sigma_1 d\tau_2 d\tau_3...d\tau_N \tag{5.13a}$$

$$\rho(\mathbf{r}_1) = \int \phi^*(\mathbf{r}_1,\mathbf{r}_2,...,\mathbf{r}_N)\phi(\mathbf{r}_1,\mathbf{r}_2,...,\mathbf{r}_N)d\sigma_1 d\tau_2 d\tau_3...d\tau_N \tag{5.13b}$$

are the reduced energy density and the spin-free diagonal of the first-order density matrix, respectively, which satisfy the conditions

$$\int \eta(\mathbf{r}_1)d\mathbf{r}_1 = <\phi|\mathcal{H}|\phi> \equiv <H> \tag{5.14a}$$

$$\int \rho(\mathbf{r}_1)d\mathbf{r}_1 = 1 \qquad \text{(if } \phi \text{ is normalized)} \tag{5.14b}$$

Use of the *RLE* is made in conjunction with the Weinstein-MacDonald energy bounds (Weinstein 1934, MacDonald 1934) as follows. First, from Eq. (5.12) and taking into account Eq. (5.14a) one obtains

$$<\varepsilon_R> = \int \rho(\mathbf{r}_1)\varepsilon_R(\mathbf{r}_1)d\mathbf{r}_1 = \int \eta(\mathbf{r}_1)d\mathbf{r}_1 \equiv <H> \tag{5.15}$$

which shows that $<\varepsilon_R>$ will be equal to the correct eigenvalue when ϕ is the correct eigenfunction. Then applying the Schwarz inequality to ϕ and $\mathcal{H}\phi$ yields

$$[\int \phi^*(r_1,r_2,...,r_N)\mathcal{H}\phi(r_1,r_2,...,r_N)d\sigma_1d\tau_2d\tau_3...d\tau_N]^2$$

$$\leq \{\int \phi^*(r_1,r_2,...,r_N)\phi(r_1,r_2,...,r_N)d\sigma_1d\tau_2d\tau_3...d\tau_N\}$$

$$\{\int [\mathcal{H}\phi(r_1,r_2,...,r_N)]^*[\mathcal{H}\phi(r_1,r_2,...,r_N)]d\sigma_1d\tau_2d\tau_3...d\tau_N\}$$

which can be rewritten as

$$\rho(r_1)\varepsilon_R^2(r_1) \leq \{\int [\mathcal{H}\phi(r_1,r_2,...,r_N)]^*[\mathcal{H}\phi(r_1,r_2,...,r_N)]d\sigma_1d\tau_2d\tau_3...d\tau_N$$

so that integration over the spatial variables of electron 1 will yield

$$<\varepsilon_R^2> \equiv \int \rho(r_1)\varepsilon_R^2(r_1)dr_1 \leq \{\int [\mathcal{H}\phi(r_1,r_2,...,r_N)]^*[\mathcal{H}\phi(r_1,r_2,...,r_N)]d\tau_1d\tau_2...d\tau_N$$

or, in short,

$$<\varepsilon_R^2> \leq <\mathcal{H}\phi|\mathcal{H}\phi> \equiv <H^2> \tag{5.16}$$

Therefore the variance for the *RLE*, in terms of the integrated quantities $<\varepsilon_R>$ and $<\varepsilon_R^2>$, may be expressed as

$$V_\varepsilon = <\rho(r_1)(\varepsilon_R - <\varepsilon_R>)^2> = <\rho(r_1)(\varepsilon_R^2 - 2\varepsilon_R<\varepsilon_R> + <\varepsilon_R>^2)>$$
$$= <\varepsilon_R^2> - <\varepsilon_R>^2 \leq <H^2> - <H>^2 \tag{5.17}$$

so that, taking into account that the corresponding variance for $H\phi$ is

$$V_\mathcal{H} = <H^2> - <H>^2 \tag{5.18}$$

one obtains

$$V_\varepsilon \leq V_\mathcal{H} \tag{5.19}$$

The Weinstein-MacDonald lower bound is given, in the present notation, by

$$\varepsilon_{WM} = <H> - [<H^2> - <H>^2]^{1/2} = <H> - V_\mathcal{H}^{1/2}$$

so that if we define a new quantity

$$\eta_R = <H> - V_\varepsilon^{1/2} \tag{5.20}$$

and take into account Eq. (5.19) we arrive at the final condition

$$\varepsilon_{WM} \leq \eta_R \leq <H>$$

It is expected that η_R will constitute a lower bound of E (Thomas *et al.* 1976).

The formulation has been generalized to energy n-density matrices (Grelland 1981) and adapted to *SCF* functions (Cohen and Frishberg 1976, Nakatsuji 1976), having been shown that the *RLE* is a constant for all *PECS* in the specific case of Hartree-Fock functions (King and Rothstein 1980). A generalization of the Cohen-Frishberg approach yields the orbital equations of MCSCF theory without use of the variational principle (Schlosser 1977).

Diverse applications of the *RLE* method have been made, such as the test of the local accuracy of *SCF* and *HF* functions (King *et al.* 1981, 1982, 1983, 1984, King and Dalke 1983, and King and Poitzsch 1984), for contour maps of the local behaviour of functions (Grelland and Almlöf 1982, King *et al.* 1983b) and in their interpretation (Casida 1985), the measure of the error of a trial function (Thomas 1983), and the test of the accuracy of probability densities (Deal 1989). More recently, locally-constrained variational calculations have been performed using a non-linear programming approach, with a set of inequality constraints, based on the reduced local energies, imposed in order to improve the local behaviour of the function (King *et al.* 1992*ab*).

5.3 Local Orbital Energies

The concept of local energy, discussed in Section 5.2.1, may be easily adapted to the *SCF* formalism. A full discussion of this topic lies outside of the scope of this work and therefore only a few comments will be presented in order to point out the relationship with other developments. In this connection the reader is referred to the specific contributions mentioned below.

Within the context of *SCF* theory the orbitals $\phi_i(\mathbf{r})$ are obtained as solutions of the eigenvalue equations

$$\mathcal{F}\phi_i(\mathbf{r}) = \varepsilon_i\phi_i(\mathbf{r})$$

where $\mathcal{F}$ is the *SCF* operator and ε_i is the corresponding orbital energy. The orbital $\phi_i(\mathbf{r})$ is expressed in terms of the spatial variables of a single electron.

The corresponding local orbital energies (Palke 1987), at a specific point $\mathbf{r}_p$, will be evaluated as

$$\varepsilon_{i(p)} = \frac{\mathcal{F}\phi_i(\mathbf{r}_p)}{\phi_i(\mathbf{r}_p)}$$

where $\phi_i(\mathbf{r}_p)$ and $\mathcal{F}\phi_i(\mathbf{r}_p)$ represent the numerical values obtained by assigning to the electron spatial variables their values at the point being considered.

At the Hartree-Fock limit it will be $\varepsilon_{i(p)} \equiv \varepsilon_i$ at each and every point. This constancy will not be observed for any other *SCF* function obtained within the framework of the expansion approximation using a limited basis set. The difference between $\varepsilon_{i(p)}$ and ε_i will then give a measure of the quality of each orbital and, therefore, of the basis set used. The corresponding variance may be defined by

$$V_i = < |\phi_i(\mathbf{r})|^2 (\varepsilon_{i(p)} - \varepsilon_i)^2 >$$

where the factor $|\phi_i(\mathbf{r})|^2$ is included in order to emphasize the most important regions of space.

The overall quality of the *SCF* function may be established (Parr and Pearson 1983, Pearson and Palke 1990) on the basis of arguments relating the orbital energy variance, the chemical potential (Parr *et al.* 1978, Politzer and Weinstein 1979, Alonso and March 1983), and the electron density (Hohenberg and Kohn 1964).

6 The Time-Dependent Schrödinger Equation

In the preceding chapters our attention has been centred on the time-independent Schrödinger equation, although time-dependent equations [such as Eqs. (1.5), (1.11), and (2.2)] have appeared in some instances.

That description will now be complemented with a development of the mathematical formulation for the time-dependence of the state functions, but the subject of measurements of physical observables will not be considered here. The conceptual interpretation of measurements is a very interesting and controversial subject and the reader is referred to the work of Wigner (1963), DeWitt and Graham (1971), Jammer (1974), d'Espagnat (1976), and Ballentine (1987), as well as to the illustrative summary of Levine (1991).

The discussion in this chapter will be centred exclusively on the time-dependent Schrödinger equation in the r-representation, taking into account the difficulties encountered in Chapter 4 for the transformation of the time-independent equation to the p-representation. [See, however, the work of Pluvinage for the simple case of the harmonic oscillator.]

6.1 The Equation and Its Solutions

There is no proper derivation of the time-dependent Schrödinger equation. Usually it is either introduced as a postulate (Houston 1959, Levine 1991) or obtained through the use of the formal transformation

$$E \rightarrow -\frac{\hbar}{i} \frac{\partial}{\partial t}$$

(Davydov 1965, Hameka 1965), as discussed when Eqs. (1.11) and (2.2) were introduced. The experimental justification of the Schrödinger equation (see, e.g., Davydov 1965) may be accepted with the usual reservations regarding inductivism (Chalmers 1991), but a discussion on this point belongs more properly to the realm of the philosophy of science. Here it will suffice to repeat that the Schrödinger equation is just a component of the formulation within the framework of relativistic quantum mechanics.

The time-dependent Schrödinger equation is

$$i\hbar \frac{\partial \Phi(\mathbf{r},t)}{\partial t} = \mathcal{H}\Phi(\mathbf{r},t) \tag{6.1}$$

where $\Phi(\mathbf{r},t)$ describes the continuous change of the system with time and $\mathcal{H}$ denotes the Hamiltonian operator, which may depend or not on the time.

In the case when the Hamiltonian operator does not depend on t, Eq. (6.1) will be written as

$$i\hbar \frac{\partial \Phi(\mathbf{r},t)}{\partial t} = \mathcal{H}(\mathbf{r})\Phi(\mathbf{r},t) \tag{6.2}$$

and a separation of variables is possible, as already discussed in Chapter 1 and Section 2.1. One can write

$$\Phi_I(\mathbf{r},t) = e^{-iE_I t/\hbar}\Psi_I(\mathbf{r}) \tag{6.3}$$

where $\Psi_I(\mathbf{r})$ is an eigenfunction

$$\mathcal{H}(\mathbf{r})\Psi_I(\mathbf{r}) = E_I\Psi_I(\mathbf{r}) \tag{6.4}$$

of the time-independent Hamiltonian operator. Substitution of Eq. (6.3) into Eq. (6.2) yields

$$i\hbar \frac{\partial \Phi_I(\mathbf{r},t)}{\partial t} = E_I e^{-iE_I t/\hbar}\,\Psi_I(\mathbf{r}) = E_I\Phi_I(\mathbf{r},t)$$

$$\mathcal{H}(\mathbf{r})\Phi_I(\mathbf{r},t) = e^{-iE_I t/\hbar}\,\mathcal{H}(\mathbf{r})\Psi_I(\mathbf{r}) = E_I\Phi_I(\mathbf{r},t)$$

for the two sides of Eq. (6.2), respectively.

The general solution, corresponding to a state without definite energy, will then be

$$\Phi(\mathbf{r},t) = \sum_I c_I\, e^{-iE_I t/\hbar}\,\Psi_I(\mathbf{r}) + \int c_E\, e^{-iEt/\hbar}\,\Psi_E(\mathbf{r})\,dE \tag{6.5}$$

where it has been assumed that the energy spectrum may consist of both discrete and continuous regions (see Section 3.1.2).

In the case when the Hamiltonian operator depends on t, we will write

$$i\hbar \frac{\partial \Phi_I(\mathbf{r},t)}{\partial t} = \mathcal{H}(\mathbf{r},t)\Phi(\mathbf{r},t) \tag{6.6}$$

with

$$\mathcal{H}(\mathbf{r},t) = \mathcal{H}(\mathbf{r}) + \mathcal{V}(\mathbf{r},t) \qquad (6.7)$$

where $\mathcal{H}(\mathbf{r})$ is the time-independent Hamiltonian operator considered above and $\mathcal{V}(\mathbf{r},t)$ represents the time-dependent potential operator (due to an external source, etc.). A separation of variables is not possible in this case and a different approach, based on the use of time-evolution operators, is required.

6.2 Time-Evolution Operators

We will define the time-evolution operator

$$\mathcal{U}(t,t_0) = e^{-i(t-t_0)\mathcal{H}(\mathbf{r})/\hbar} \qquad (6.8)$$

which has the properties

$$\mathcal{U}(t_0,t_0) = 1 \qquad\qquad \mathcal{U}(t,t')\mathcal{U}(t',t_0) = \mathcal{U}(t,t_0)$$

$$\mathcal{U}(t,t_0) = \mathcal{U}^{-1}(t_0,t) \qquad\qquad \mathcal{U}^{-1}(t,t_0)\mathcal{U}(t,t_0) = 1$$

$$[\mathcal{H}(\mathbf{r}),\mathcal{U}(t,t_0)] = 0 \qquad\qquad [\mathcal{H}(\mathbf{r},t),\mathcal{U}(t,t_0)] = 0$$

and satisfies the evolution equation

$$i\hbar\,\frac{\partial\mathcal{U}(t,t_0)}{\partial t} = \mathcal{H}(\mathbf{r})\mathcal{U}(t,t_0) \qquad (6.9)$$

If $\mathcal{H}(\mathbf{r})$ is Hermitian, which is the case considered here, then $\mathcal{U}(t,t_0)$ is a unitary operator. In such a case, given the transformation

$$\psi' = \mathcal{U}(t,t_0)\psi$$

it is

$$\psi'^* = \psi'^*\mathcal{U}^{-1}(t,t_0)$$

so that the norm

$$\langle\psi'|\psi'\rangle = \langle\psi|\mathcal{U}^{-1}(t,t_0)\mathcal{U}(t,t_0)|\psi\rangle = \langle\psi|\psi\rangle$$

is conserved.

6.3 Time-Evolution Representations

The time-evolution operator may be used in order to express the evolution of the state functions and/or the operators corresponding to physical observables. The three possible approaches are embodied, respectively, in the Schrödinger, Heisenberg, and Dirac (interaction) representations. The Schrödinger and Heisenberg pictures are used in the case when the Hamiltonian operator is independent of t while the interaction picture was developed for the case when the Hamiltonian operator depends on t. These three pictures will be examined below, with particular attention given to the interaction picture. [For additional information, the reader is referred to the work of Davydov (1965), Akhiezer and Berestetskii (1965), Klauder and Sudarshan (1968), Bialynicki-Birula and Bialynicka-Birula (1975), Loudon (1983), and Gomes and Paniagua (1992), among others, but a word of caution is appropriate regarding the diverse notations.]

6.3.1 The Schrödinger Representation

The time-dependent Schrödinger equation, Eq. (6.2), represents the basic equation. The evolution of the functions is expressed, in general, as

$$\Phi(\mathbf{r},t) = \mathcal{U}(t,t_0)\Phi(\mathbf{r},t_0) \tag{6.10}$$

which reduces, when $t_0 = 0$, to

$$\Phi(\mathbf{r},t) = \mathcal{U}(t,0)\Phi(\mathbf{r},0) \tag{6.11}$$

where $\Phi(\mathbf{r},0)$ is given by Eq. (6.5). For a particular state function one has, in a similar way,

$$\Phi_I(\mathbf{r},t) = \mathcal{U}(t,t_0)\Phi_I(\mathbf{r},t_0) \tag{6.12}$$

and

$$\Phi_I(\mathbf{r},t) = \mathcal{U}(t,0)\Phi_I(\mathbf{r},0) \equiv \mathcal{U}(t,0)\Psi_I(\mathbf{r}) \tag{6.13}$$

A series expansion of $\mathcal{U}(t,0)$ allows us to relate the present formulation with the one given in Section 6.1. For example, for Eq. (6.13) we can write

$$\Phi_I(\mathbf{r},t) = \{ \sum_{k=0}^{\infty} \frac{1}{k!} [-it\mathcal{H}(\mathbf{r})/\hbar]^k \}\Psi_I(\mathbf{r}) = \{ \sum_{k=0}^{\infty} \frac{1}{k!} [-itE_I/\hbar]^k \}\Psi_I(\mathbf{r}) = e^{-iE_I t/\hbar}\Psi_I(\mathbf{r})$$

which is the expression in Eq. (6.3).

Within the context of this representation, the operators $P(r)$ corresponding to physical observables are maintained independent of t,

$$i\hbar \frac{\partial P(r)}{\partial t} = 0$$

and the time-evolution of their expectation values is then given by

$$< \Phi_I(r,t)|P(r)\Phi_J(r,t) > = e^{i(E_I - E_J)t/\hbar} < \Psi_I(r)|P(r)\Psi_J(r) >$$

6.3.2 The Heisenberg Representation

In this picture the functions are maintained independent of t,

$$i\hbar \frac{\partial \Psi_I(r)}{\partial t} = 0$$

and the time evolution is introduced in the operator through the transformation

$$P(r,t) = U^{-1}(t,t_0)P(r)U(t,t_0)$$

so that the corresponding evolution equation is

$$i\hbar \frac{\partial P(r,t)}{\partial t} = - \mathcal{H}(r)U^{-1}(t,t_0)P(r)U(t,t_0) + U^{-1}(t,t_0)P(r)U(t,t_0)\mathcal{H}(r) = [P(r,t),\mathcal{H}(r)]]$$

where use has been made of the fact the $\mathcal{H}(r)$ and $U(t,t_0)$ commute.

The expectation values of these operators are then

$$<\Psi_I(r)|P(r,t)\Psi_J(r)> = <\Psi_I(r)|U^{-1}(t,t_0)P(r)U(t,t_0)\Psi_J(r)>$$

$$= <\Phi_I(r,t)|P(r)\Phi_J(r,t)> = e^{i(E_I - E_J)t/\hbar} <\Psi_I(r)|P(r)\Psi_J(r)>$$

which is the result obtained in the Schrödinger representation.

The Hamiltonian operator in this representation

$$U^{-1}(t,t_0)\mathcal{H}(r)U(t,t_0) = U^{-1}(t,t_0)U(t,t_0)\mathcal{H}(r) = \mathcal{H}(r)$$

remain unchanged.

6.3.3 The Interaction Representation

This picture represents an intermediate between the Schrödinger and the Heisenberg pictures and it is based on the use of new functions, denoted as *interaction functions*. For simplicity in the notation, in order to avoid an additional subscript, these new functions will be denoted by $\phi(\mathbf{r},t)$.

The interaction functions are defined as

$$\phi(\mathbf{r},t) = \mathcal{U}^{-1}(t,\tau)\Phi(\mathbf{r},t) \tag{6.14}$$

where $\Phi(\mathbf{r},t)$ is the function in the Schrödinger representation and τ denotes the time when the Schrödinger and the interaction pictures coincide. [If desired, τ may be set arbitrarily to 0.]

The corresponding evolution equation is then

$$i\hbar\,\frac{\partial\phi(\mathbf{r},t)}{\partial t} = -\mathcal{H}(r)\mathcal{U}^{-1}(t,\tau)\Phi(\mathbf{r},t) + \mathcal{U}^{-1}(t,\tau)\mathcal{H}(\mathbf{r},t)\Phi(\mathbf{r},t)$$

taking into account Eq. (6.2). This equation may be rewritten as

$$i\hbar\,\frac{\partial\phi(\mathbf{r},t)}{\partial t} = -\mathcal{H}(\mathbf{r})\phi(\mathbf{r},t) + \mathcal{U}^{-1}(t,\tau)\mathcal{H}(\mathbf{r},t)\mathcal{U}(t,\tau)\mathcal{U}^{-1}(t,\tau)\Phi(\mathbf{r},t)$$

$$= -\mathcal{H}(\mathbf{r})\phi(\mathbf{r},t) + \mathcal{U}^{-1}(t,\tau)\{\mathcal{H}(\mathbf{r}) + \mathcal{V}(\mathbf{r},t)\}\mathcal{U}(t,\tau)\phi(\mathbf{r},t) \tag{6.15}$$

$$= \mathcal{U}^{-1}(t,\tau)\mathcal{V}(\mathbf{r},t)\mathcal{U}(t,\tau)\phi(\mathbf{r},t)$$

and finally as

$$i\hbar\,\frac{\partial\phi(\mathbf{r},t)}{\partial t} = \hbar(\mathbf{r},t)\phi(\mathbf{r},t) \tag{6.16}$$

which is the time-dependent Schrödinger equation for the interaction functions and where

$$\hbar(\mathbf{r},t) = \mathcal{U}^{-1}(t,\tau)\mathcal{V}(\mathbf{r},t)\mathcal{U}(t,\tau) \tag{6.17}$$

represents the Hamiltonian operator in the interaction picture.

The evolution of the interaction functions may be expressed also as

$$\phi(\mathbf{r},t) = \mathbf{u}(t,t_0)\phi(\mathbf{r},t_0) \tag{6.18}$$

where $\mathbf{u}(t,t_0)$ is the interaction time-evolution operator, with the corresponding

$$i\hbar \frac{\partial u(t,t_0)}{\partial t} = h(\mathbf{r},t)u(t,t_0) \tag{6.19}$$

evolution equation, which is obtained from Eqs. (6.16) and (6.18).

Substituting into Eq. (6.18) the expressions of $\phi(\mathbf{r},t)$ and $\phi(\mathbf{r},t_0)$ in terms of $\Phi(\mathbf{r},t)$ and $\Phi(\mathbf{r},t_0)$, respectively, according to Eq. (6.14), one obtains

$$u^{-1}(t,\tau)\Phi(\mathbf{r},t) = u(t,t_0)u^{-1}(t_0,\tau)\Phi(\mathbf{r},t_0)$$

which can be rewritten as

$$\Phi(\mathbf{r},t) = u(t,\tau)u(t,t_0)u^{-1}(t_0,\tau)\Phi(\mathbf{r},t_0) \tag{6.20}$$

after multiplication on the left by $u(t,\tau)$. This expression shows that the evolution of the functions $\Phi(\mathbf{r},t)$, when taking into account the existence of the term $V(\mathbf{r},t)$, is governed by the time-evolution operator

$$u_I(t,t_0) = u(t,\tau)u(t,t_0)u^{-1}(t_0,\tau) \tag{6.21}$$

so that the solution of the problem requires the knowledge of $u(t,t_0)$.

6.4 The Solution of the Interaction Evolution Equation

6.4.1 The Perturbation Series Expansion

The integral equation

$$i\hbar\{\phi(\mathbf{r},t) - \phi(\mathbf{r},t_0)\} = \int_{t_0}^{t} dt'\, h(\mathbf{r},t')\phi(\mathbf{r},t')$$

equivalent to Eq. (6.16), has (see below) the formal solution

$$\phi(\mathbf{r},t) = \phi(\mathbf{r},t_0)\, T \exp\{ -\frac{i}{\hbar} \int_{t_0}^{t} dt'\, h(\mathbf{r},t')\}$$

where T is the Dyson time-ordering (chronological) operator (Dyson 1949), which orders a product of time-dependent operators [see also the work of Pechukas and Light (1966), Klauder and Sudarshan (1968), and Galindo and Pascual (1991)]. Therefore, taking into account Eq. (6.18), one obtains

$$\mathbf{u}(t,t_0) = T \exp\{ -\frac{i}{\hbar} \int_{t_0}^{t} dt' \, \mathbf{h}(\mathbf{r},t')\} \tag{6.22}$$

and

$$\mathbf{u}_1(t,t_0) = \mathbf{u}(t,\tau) \, T \exp\{ -\frac{i}{\hbar} \int_{t_0}^{t} dt' \, \mathbf{h}(\mathbf{r},t')\} \mathbf{u}^{-1}(t_0,\tau) \tag{6.23}$$

Alternatively, one can proceed directly from the evolution equation, Eq. (6.19), for the interaction time-evolution operator. Its equivalent integral equation

$$i\hbar\{\mathbf{u}(t,t_0) - \mathbf{u}(t_0,t_0)\} = i\hbar\{\mathbf{u}(t,t_0) - 1\} = T \int_{t_0}^{t} dt' \, \mathbf{h}(\mathbf{r},t')\mathbf{u}(t',t_0)$$

(where T denotes again the Dyson chronological operator) may be rewritten as

$$\mathbf{u}(t,t_0) = 1 + \frac{T}{i\hbar} \int_{t_0}^{t} dt' \, \mathbf{h}(\mathbf{r},t')\mathbf{u}(t',t_0)$$

Iteration of this equation yields [see e.g., Pechukas and Light (1966)]

$$\mathbf{u}(t,t_0) = 1 + \frac{1}{i\hbar} \int_{t_0}^{t} dt_1 \, \mathbf{h}(\mathbf{r},t_1) + \left(\frac{1}{i\hbar}\right)^2 \int_{t_0}^{t} dt_2 \int_{t_0}^{t_2} dt_1 \, \mathbf{h}(\mathbf{r},t_2)\mathbf{h}(\mathbf{r},t_1) + \dots$$

$$= 1 + \frac{1}{i\hbar} \int_{t_0}^{t} dt_1 \, \mathbf{u}^{-1}(t_1,\tau)\mathbf{v}(\mathbf{r},t_1)\mathbf{u}(t_1,\tau)$$

$$+ \left(\frac{1}{i\hbar}\right)^2 \int_{t_0}^{t} dt_2 \int_{t_0}^{t_2} dt_1 \, \mathbf{u}^{-1}(t_2,\tau)\mathbf{v}(\mathbf{r},t_2)\mathbf{u}(t_2,\tau)\mathbf{u}^{-1}(t_1,\tau)\mathbf{v}(\mathbf{r},t_1)\mathbf{u}(t_1,\tau) + \dots$$

$$= 1 + \frac{1}{i\hbar} \int_{t_0}^{t} dt_1 \, \mathbf{u}^{-1}(t_1,\tau)\mathbf{v}(\mathbf{r},t)\mathbf{u}(t_1,\tau)$$

$$+ \left(\frac{1}{i\hbar}\right)^2 \int_{t_0}^{t} dt_2 \int_{t_0}^{t_2} dt_1 \, \mathbf{u}^{-1}(t_2,\tau)\mathbf{v}(\mathbf{r},t_2)\mathbf{u}(t_2,\tau)\mathbf{u}(\tau,t_1)\mathbf{v}(\mathbf{r},t_1)\mathbf{u}(t_1,\tau) + \dots$$

$$= 1 + \frac{1}{i\hbar} \int_{t_0}^{t} dt_1 \, \mathbf{u}^{-1}(t_1,\tau)\mathbf{v}(\mathbf{r},t_1)\mathbf{u}(t_1,\tau) \tag{6.24}$$

$$+ \left(\frac{1}{i\hbar}\right)^2 \int_{t_0}^{t} dt_2 \int_{t_0}^{t_2} dt_1 \, \mathbf{u}^{-1}(t_2,\tau)\mathbf{v}(\mathbf{r},t_2)\mathbf{u}(t_2,t_1)\mathbf{v}(\mathbf{r},t_1)\mathbf{u}(t_1,\tau) + \dots$$

Equation (6.22) and the symbolic interpretation of its exponential (Feynman 1951) represent a shorthand representation of this equation. [See also the work of Birkhoff (1937), Kubo (1952), and Fer (1958ab) for other approaches to this problem.]

Combining Eqs. (6.21) and (6.24) yields finally

$$u_I(t,t_0) = u(t,t_0) + \frac{1}{i\hbar} \int_{t_0}^{t} dt_1\, u(t,\tau)u^{-1}(t_1,\tau)\nu(r,t_1)u(t_1,\tau)u^{-1}(t_0,\tau)$$

$$+ \left(\frac{1}{i\hbar}\right)^2 \int_{t_0}^{t} dt_2 \int_{t_0}^{t_2} dt_1 u(t,\tau)u^{-1}(t_2,\tau)\nu(r,t_2)u(t_2,t_1)\nu(r,t_1)u(t_1,\tau)u^{-1}(t_0,\tau) + \dots$$

$$= u(t,t_0) + \frac{1}{i\hbar} \int_{t_0}^{t} dt_1 u(t,t_1)\nu(r,t_1)u(t_1,t_0)$$

$$+ \left(\frac{1}{i\hbar}\right)^2 \int_{t_0}^{t} dt_2 \int_{t_0}^{t_2} dt_1 u(t,t_2)\nu(r,t_2)u(t_2,t_1)\nu(r,t_1)u(t_1,t_0) + \dots$$

$$= u(t,t_0) + \sum_{n=1}^{\infty} u_n(t,t_0) \tag{6.25a}$$

where

$$u_n(t,t_0) = \left(\frac{1}{i\hbar}\right)^n \int_{t_0}^{t} dt_n \int_{t_0}^{t_n} dt_{n-1} \dots \int_{t_0}^{t_2} dt_1\, w(t,t_n,t_{n-1}, \dots t_1,t_0) \tag{6.25b}$$

and

$$w(t,t_n,t_{n-1},\dots t_1,t_0) = u(t,t_n)\nu(r,t_n)u(t_n,t_{n-1})\nu(r,t_{n-1})u(t_{n-1},t_{n-2})$$

$$\dots\dots u(t_2,t_1)\nu(r,t_1)u(t_1,t_0) \tag{6.25c}$$

6.4.2 The Magnus Formula

Truncation of the above series expansion, Eq. (6.25), determines that the approximation obtained for the operator $u(t,t_0)$ will not be unitary and will only be accurate for small time intervals, $(t-t_0)$, or for small interactions. Therefore alternate approaches must be considered, especially those yielding an exponential form for the time-evolution operator, which is appropriate for unitary transformations. Most of the work has been carried out within the context of Lie algebra and mention must be made, in particular, of the work of Magnus (1954), Wichman (1961), Wei (1963), Wei and Norman (1963a), Pechukas and Light (1966), Alhassid and Levine (1978), Shin (1983), Gazdy and Micha (1985), Benjamin (1986), Ciacci et al. (1986), Wolf and Korsch (1988), Fernandez (1987, 1988, 1989, 1990), and Fernandez et al. (1987, 1990). [In addition, the reader is referred to the work of Bourbaki (1961), Jacobson (1962), Mathews and Walker (1964), Humphreys (1972), Gilmore (1974), Wybourne (1974), Sanchez Mondragon and Wolf (1986), and Sattinger and Weaver (1986), among others, for details and applications of Lie algebra.] Here we will examine some of the basic ideas, as embodied in the work of Wei and Norman (1963) and Magnus (1954), respectively.

In many instances, the time-dependent component of the Hamiltonian operator may be written as a finite sum

$$\mathcal{V}(r,t) = \sum_{j=1}^{m} v_j(t)x_j$$

where the $v_j(t)$ form a set of linearly-independent, complex-valued, time-dependent functions and the x_j are constant operators; the set $\{x_j\}$ may be expanded by repeated commutation to a Lie algebra L. If L is finite-dimensional, the evolution operator may be uncoupled (Wei and Norman 1963a) into

$$u(t,t_0) = u_1(t,t_0)u_2(t,t_0) \dots u_n(t,t_0)$$

where n (usually $n \geq m$) is the number of linearly-independent operators x_j in L, with each $u_j(t,t_0)$ satisfying an evolution equation

$$\frac{\partial u_j(t,t_0)}{\partial t} = \dot{g}_j(t)x_j u_j(t,t_0) \tag{6.26}$$

where the $g_j(t)$ are complex-valued, time-dependent functions and the notation $\dot{g}_J(t)$ denotes differentiation with respect to t, as usual. The solutions of Eqs (6.26) are then

$$u_j(t,t_0) = e^{g_j(t)x_j} \qquad \text{with } u_j(t_0,t_0) = 1$$

and therefore

$$u(t,t_0) = \prod_{j=1}^{n} e^{g_j(t)x_j} \tag{6.27}$$

which is the exact solution. The functions $g_j(t)$ are the solutions of a set of non-linear differential equations

$$\dot{g}_j(t) = \sum_{k=1}^{n} v_k(t)G_{kj}(t)$$

where the $G_{kj}(t)$ are non-linear functions of the $g_j(t)$. In the case when L is a solvable Lie algebra, the $G_{kj}(t)$ form a triangular matrix and the set of equations may be solved by quatrature (Wei and Norman 1963b). In general, however, Eq. (6.27) is not global and its range of applicability may depend on the basis and its order (Fernandez 1989).

Magnus (1954) showed that the time-evolutions operator can be expressed locally as

$$u(t,t_0) = \exp \left\{ \sum_{j=1}^{n} f_j(t,t_0)x_j \right\}$$

where the $f_j(t,t_0)$ form a set of scalar, time-dependent functions, which satisfy a set of non-linear differential equations (see below). This approach is based on expressing $u(t,t_0)$ as

$$u(t,t_0) = \exp\{A(t,t_0)\}$$

where $A(t,t_0)$ is an anti-Hermitian operator. Magnus (1954) expressed the operator $A(t,t_0)$ as an infinite series, in which the n-th term is a sum of integrals of n-fold multiple commutators of $h(r,t)$. Each term in the series is anti-Hermitian and therefore a unitary approximation to $u(t,t_0)$ is obtained even when the series is truncated.

Expressing $A(t,t_0)$ as

$$A(t,t_0) = A_1(t,t_0) + A_2(t,t_0) + ...$$

the first four terms of the Magnus expansion (Magnus 1964, Robinson 1963, Pechukas and Light 1966) are

$$A(t,t_0) = \int_{t_0}^{t} dt_1 \chi(t_1)$$

$$A_2(t,t_0) = -\frac{1}{2} \int_{t_0}^{t} dt_2 \int_{t_0}^{t_2} dt_1 [t_1,t_2]$$

$$A_3(t,t_0) = -\frac{1}{6} \int_{t_0}^{t} dt_3 \int_{t_0}^{t_3} dt_2 \int_{t_0}^{t_2} dt_1 \{[t_1,[t_2,t_3]] + [[t_1,t_2],t_3]\}$$

$$A_4(t,t_0) = -\frac{1}{12} \int_{t_0}^{t} dt_4 \int_{t_0}^{t_4} dt_3 \int_{t_0}^{t_3} dt_2 \int_{t_0}^{t_2} dt_1 \{[t_1,[[t_2,t_3],t_4]]$$

$$+ [[t_1,[t_2,t_3]],t_4] + [[t_1,t_2],[t_3,t_4]]\}$$

where

$$[t_j,t_k] = [\chi(t_j),\chi(t_k)] \qquad \qquad \chi(t_j) = -\frac{i}{\hbar} h(r,t)$$

The Magnus expansion may not be applicable under certain conditions (Maricq 1982, Fel'dman 1984). It converges faster than the perturbation series when the interval (t-t$_0$) is small but its general convergence properties are difficult to prove and for that reason simple models have been investigated (Wei and Norman 1963, Fel'dman 1984, Salzman 1985, 1986ab, 1987, and Fernandez 1987) in order to obtain additional information. This expansion has found applications in the analysis of magnetic resonance experiments (Maricq 1982, 1986, Fel'dman 1984, and Salzman 1986ab, 1987) and in the study of multiple Coulomb excitation of deformed nuclei (Robinson 1963) and of the vibrational energy transfer in collisions between diatomic molecules (Kelley 1972).

Fernandez (1989) has presented an alternate formulation involving the solution of linear differential equations.

6.5 Evolution of the System: Transition Probabilities

The preceding formulation allows us to follow the evolution of the system under the influence of a source of interaction, which varies with time. For example, the system may be constituted by two (atomic or molecular) subsystems, initially at an infinite separation from each other. The interaction comes into effect when their separation (and relative orientation, if applicable) changes.

Let us assume that the system is described, at $t = t_0$, by the function $\Psi_J(\mathbf{r})$. The interaction is then switched on and the system will evolve according to

$$\Phi_J(\mathbf{r},t) = \mathcal{U}_I(t,t_0)\Psi_J(\mathbf{r})$$

and $\Phi_J(\mathbf{r},t)$ will have contributions from all the states of the free system. [For simplicity and without loss of generality we will consider in the discussion only the discrete spectrum of the free system.]

The weight of the K-th state in $\Phi_J(\mathbf{r},t)$ is given by

$$w_{KJ} = \left| < \Psi_K(\mathbf{r})|\Phi_J(\mathbf{r},t) > \right|^2 \tag{6.28}$$

and gives a measure of the transition from the J-th state to the K-th state under the influence of the interaction. When proceeding within the context of the perturbation expansion approximation for $\mathcal{U}_I(t,t_0)$ we must then evaluate

$$< \Psi_K(\mathbf{r})|\{\mathcal{U}(t,t_0) + \sum_{n=1}^{\infty} \mathcal{U}_n(t,t_0)\}\Psi_J(\mathbf{r}) > \tag{6.29}$$

using Eqs. (6.25). With the notation

$$\mathcal{V}(r,t_n)\Psi_J(r) = \sum_{K_n} \Psi_{K_n}(r)V(K_n,J;t_n)$$

$$V(K_n,J;t_n) = \langle\Psi_{K_n}(r)|\mathcal{V}(r,t_n)\Psi_J(r)\rangle$$

and taking into account that

$$\mathcal{U}(t_1,t_0)\Psi_J(r) = e^{-iE_J(t_1-t_0)/\hbar}\Psi_J(r)$$

$$\mathcal{V}(r,t_1)\mathcal{U}(t_1,t_0)\Psi_J(r) = \mathcal{V}(r,t_1)\Psi_J(r)\,e^{-iE_J(t_1-t_0)/\hbar} = \{\sum_{K_1} \Psi_{K_1}(r)V(K_1,J;t_1)\}e^{-iE_J(t_1-t_0)/\hbar}$$

$$\mathcal{U}(t_2,t_1)\mathcal{V}(r,t_1)\mathcal{U}(t_1,t_0)\Psi_J(r)$$

$$= \mathcal{U}(t_2,t_1)\{\sum_{K_1} \Psi_{K_1}(r)V(K_1,J;t_1)\}e^{-iE_J(t_1-t_0)/\hbar}$$

$$= \{\sum_{K_1} e^{-iE_{K_1}(t_2-t_1)/\hbar}\Psi_{K_1}(r)V(K_1,J;t_1)\}e^{-iE_J(t_1-t_0)/\hbar}$$

. . .

one finally obtains

$$\langle\Psi_K(r)|w(t,t_n,t_{n-1}, \dots\, t_1,t_0)\Psi_J(r)\rangle$$

$$= \sum_{K_{n-1}} \sum_{K_{n-2}} \dots \sum_{K_1} e^{-iE_K(t-t_n)/\hbar}V(K,K_{n-1};t_n)$$

$$e^{-iE_{K_{n-1}}(t_n-t_{n-1})/\hbar}V(K_{n-1},K_{n-2};t_{n-1})$$

. . .

$$e^{-iE_{K_1}(t_2-t_1)/\hbar}V(K_1,J;t_1)\,e^{-iE_J(t_1-t_0)/\hbar}$$

Substitution into Eq. (6.29), taking into account Eqs. (6.25), and then into Eq. (6.28) will yield the final expression for the transition probability, which is omitted here for simplicity, given its complicated nature.

Examples of the equivalent formulation, using the Magnus formula, are given by Pechukas and Light (1966).

The Non-Linear Schrödinger Equation

A century and a half ago, chance opened up an extraordinary field of research, which has shown an exponential growth during our lifetime. In terms of the knowledge achieved of physical and chemical phenomena, the observation of solitary waves in a Scottish canal by Russell and the visualization by Kekule of aromatic rings in the flickering flames in a fireplace are comparable. In both instances it would have been difficult, if not impossible, to foresee what lay ahead, as evidenced in particular by the amazing progression from solitary waves to solitons in so many areas of physics. At present, however, the necessary sophistication of research, in order to study more complex phenomena, makes a physical visualization rather improbable and work now proceeds through elegant mathematical formulations and powerful computational techniques.

The following chapter on the non-linear Schrödinger equation is included for completeness, even though that equation is not found within the context of quantum mechanics. Justice is not done to the subject, particularly insofar as solitons are concerned, but the topics of space and time discretization establish a connection with the techniques for docking procedures in drug research and for the molecular dynamics of proteins.

7 The Non-Linear Schrödinger Equation

7.1 Solitary Waves and Solitons

The observation of a rather peculiar type of wave in the Union canal in Scotland stimulated Russell (1844) to a study of waves, which resulted in their classification as gregarious (oscillatory water waves and wave groups, capillary surface waves) and solitary waves (waves of translation, corpuscular waves). The designation *solitary wave*, in particular, applies to any plane wave which translates in space without change of shape.

The Korteweg-de Vries equation, usually abbreviated as the *KdV* equation (Korteweg and de Vries 1895),

$$ -\frac{\partial \Phi}{\partial t} = \frac{\partial^3 \Phi}{\partial x^3} + \Phi \frac{\partial \Phi}{\partial x} \qquad \text{or} \qquad -\Phi_t = \Phi_{xxx} + \Phi \Phi_x $$

(where the standard notation in this field has been used in the second expression) exhibits solitary wave solutions. The study of the *KdV* equation by Zabusky and Kruskal (1965) showed, for some initial conditions, the existence of pulses resembling the solitary wave solution, with the characteristic that the larger pulses overtook the smaller ones, with a non-linear interaction, and emerged with a phase shift but retaining their shape (height and width) and speed. This type of solitary wave was denoted as *soliton*. [See also the reviews of Zabusky (1967), Kruskal (1975), and Miura (1976).]

The *KdV* equation is an example of a non-linear, dispersive partial differential equation. Other examples of this type of evolution equation, which also exhibit soliton solutions, are the non-linear Schrödinger equation (hereafter abbreviated as the *NLS* equation) and the sine-Gordon equation. Solitons result from a balance between dispersion (which spreads the wave) and the non-linearity (which steepens the wave) but they are not a characteristic of all the non-linear, dispersive equations.

The widespread significance of solitons has provoked an intensive effort in their study. As our interest here is restricted to the *NLS* equation, the reader interested in solitons in general is referred to the abundant literature on the subject. Just as a sample, without trying to be exhaustive, mention will be made of the work of Lonngren and Scott (1978), Bullough and Caudrey (1980), Eilenberger (1981), Rebbi and Solani (1984), Davydov (1985,1991), Takeno (1985), Faddeev and Takhtajan (1987), Fordy (1990), Olver and Sattinger (1990), Taylor (1992), and Abdullaev (1994). Particular attention should be given to the work of Rebbi and Solani (1984), which includes a collection of original papers on the subject.

7.2 The 1-Dimensional *NLS* Equation

The *NLS* equation appears in connection with, or is used as a model in, the study of such diverse problems as non-linear optics, superconductivity, vortex motion, and plasma/low temperature/condensed matter physics (see also Chapter 1). Taking into account the historical development of solitary waves, it should also be noted that the *NLS* equation appears in the study of the evolution of deep water waves (Yuen and Lake 1975, 1978).

The one-dimensional *NLS* equation is

$$-\,\mathrm{i}\,\frac{\partial \Phi}{\partial t} = \frac{\partial^2 \Phi}{\partial x^2} + q|\Phi|^2\Phi \qquad \text{or} \qquad -\,\mathrm{i}\,\Phi_t = \Phi_{xx} + q|\Phi|^2\Phi \qquad\qquad (7.1)$$

where $\Phi(x,t)$ is a complex-valued function in the domain $-\infty < x < \infty$, $t > 0$, and q is a real-valued parameter; $|\Phi|^2$ stands for $\Phi^*\Phi$. This equation describes the evolution of any weakly non-linear, strongly dispersive almost monochromatic wave. The designation *cubic NLS* is often used for this equation in order to indicate that the non-linear term is of the form $|\Phi|^2\Phi$.

The *NLS* equation also exhibits soliton solutions (for appropriate initial conditions) for $q = 2N^2$, $N = 1,2,...$ The solitons may propagate in both directions, with speeds and amplitudes which are independent of each other. As the *NLS* equation allows for waves of different speeds, there is the possibility of collision of solitons, with the type of collision depending on the ratio of the speeds of the two solitons: in a collision as particles there is transfer of momentum and the solitons change direction while in a collision as waves the solitons pass through each other unchanged (except for a phase shift).

The *NLS* equation becomes, for q = 0,

$$\mathrm{i}\,\frac{\partial \Phi}{\partial t} = -\frac{\partial^2 \Phi}{\partial x^2} \qquad\qquad (7.2)$$

which is the time-dependent linear Schrödinger equation (see Eq. 6.1) for a free-particle with m = 1/2 (in atomic units).

7.2.1 Conservation Laws and Conserved Quantities

Two reasons dictate that, before examining the analytic and numerical solution of the *NLS* equation, we should discuss the local conservation laws and the related conserved quantities. On one hand the existence of both conservation laws and soliton solutions may be related (Miura 1978) and, in addition, the conserved quantities may play a role in the numerical solution of the equation.

A polynomial conservation law is defined (Herman 1978, Fordy 1990) by a pair of polynomial functions, $T(u, u_x, u_{xx}, ...)$ and $X(u, u_x, u_{xx}, ...)$, such that

$$\frac{\partial T}{\partial t} = -\frac{\partial X}{\partial x} \qquad \text{or} \qquad T_t + X_x = 0$$

With appropriate boundary conditions, the local conservation laws will lead to conserved quantities. For the *KdV* equation, in particular, the existence of an infinite sequence of conserved quantities has been discussed by Miura *et al.* (1968) but here we will restrict our attention to the quantities

$$N = \int_{-\infty}^{\infty} |\Phi|^2 dx \tag{7.3a}$$

$$H = \int_{-\infty}^{\infty} \{ \frac{1}{2} |\Phi_x|^2 - \frac{1}{4} q \, |\Phi|^4 \} dx \tag{7.3b}$$

$$P = \frac{1}{2i} \int_{-\infty}^{\infty} \{ \Phi^* \Phi_x - \Phi_x^* \Phi \} dx \tag{7.3c}$$

denoted, respectively, as the charge (or number of particles), the Hamiltonian (or energy integral), and the momentum. [See, e.g., Eilenberger (1981), Faddeev and Takhtajan (1987), and Abdullaev (1994).] The soliton has a finite energy and the integrals of motion, N and P, are finite.

Diverse approaches exist for the determination of the expressions of the conserved quantities (see, e.g., Miura *et al.* 1968, Zakharov and Shabat 1972, Herman 1978, Miura 1978, Faddeev 1980, Fordy 1990) but here we will examine a simple procedure.

Multiplication of Eq. (7.1) on the left with Φ^* and integration over all space, with subsequent manipulation, yields

$$\int_{-\infty}^{\infty} \Phi^*(i\Phi_t + \Phi_{xx} + q \, |\Phi|^2\Phi)dx$$

$$= \int_{-\infty}^{\infty} \{i \, (\Phi^*\Phi_t + \Phi_t^* \, \Phi) - i\Phi_t^*\Phi + \Phi^*\Phi_{xx} + q|\Phi|^2\Phi^*\Phi\}dx$$

$$= \int_{-\infty}^{\infty} \{i \, \frac{\partial}{\partial t}(\Phi^*\Phi) - (\Phi_{xx}^* + q|\Phi|^2\Phi^*)\Phi + \Phi^*\Phi_{xx} + q|\Phi|^2\Phi^*\Phi\}dx$$

$$= \int_{-\infty}^{\infty} \{i \, \frac{\partial}{\partial t}(\Phi^*\Phi) + \frac{\partial}{\partial x}(\Phi^*\Phi_x - \Phi_x^*\Phi)\}dx$$

$$= i \, \frac{\partial}{\partial t} \int_{-\infty}^{\infty} |\Phi|^2 dx + [\Phi^*\Phi_x - \Phi_x^*\Phi]_{-\infty}^{\infty} = 0$$

where the last term vanishes if Φ vanishes rapidly as $x \to \pm \infty$. Therefore the quantity N, Eq. (7.3a), is invariant.

Similarly one can write

$$\int_{-\infty}^{\infty} \{\Phi_t^*(i\Phi_t + \Phi_{xx} + q \, |\Phi|^2\Phi) + (-i\Phi_t^* + \Phi_{xx}^* + q \, |\Phi|^2\Phi^*)\Phi_t\}dx$$

$$= \int_{-\infty}^{\infty} \{(\Phi_t^*\Phi_{xx} + \Phi_{xx}^* \, \Phi_t) + q \, |\Phi|^2(\Phi_t^*\Phi + \Phi^*\Phi_t)\}dx$$

$$= \int_{-\infty}^{\infty} \{ \frac{\partial}{\partial x}(\Phi_t^*\Phi_x + \Phi_x^*\Phi_t) - (\Phi_{xt}^*\Phi_x + \Phi_x^*\Phi_{xt}) + q|\Phi|^2(\Phi_t^*\Phi + \Phi^*\Phi_t)\}dx$$

$$= [\Phi_t^*\Phi_x + \Phi_x^*\Phi_t]_{-\infty}^{\infty} - \frac{\partial}{\partial t} \int_{-\infty}^{\infty} \{\Phi_x^*\Phi_x - \frac{1}{2} \, q|\Phi|^4\}dx = 0$$

where Φ_{xt} denotes the derivative of Φ with respect to x and t. The first term in this equation vanishes if Φ vanishes rapidly as $x \to \pm\infty$ and therefore the quantity H, Eq. (7.3b), is invariant.

Regarding the invariance of the momentum, one can write

$$\frac{\partial}{\partial t} \int_{-\infty}^{\infty} (\Phi^*\Phi_x - \Phi_x^*\Phi)dx = \int_{-\infty}^{\infty} (\Phi_t^*\Phi_x + \Phi^*\Phi_{xt} - \Phi_{xt}^*\Phi - \Phi_x^*\Phi_t)dx$$

$$= i \int_{-\infty}^{\infty} \{-\Phi_{xx}^*\Phi_x - \Phi_x^*\Phi_{xx} + \Phi^*\Phi_{xxx} + \Phi_{xxx}^*\Phi + 2q \, |\Phi|^2(\Phi^*\Phi_x + \Phi_x^*\Phi))\}dx$$

$$= i \int_{-\infty}^{\infty} \frac{\partial}{\partial x} \{\Phi_{xx}^*\Phi - 2\Phi_x^*\Phi_x + \Phi^*\Phi_{xx} + q|\Phi|^4\}dx$$

$$= i \, [\Phi_{xx}^*\Phi - 2\Phi_x^*\Phi_x + \Phi^*\Phi_{xx} + q|\Phi|^4]_{-\infty}^{\infty} = 0$$

if Φ vanishes rapidly as $x \to \pm \infty$.

7.2.2 Analytic Solution

Zakharov and Shabat (1972) solved Eq. (7.1) exactly using the inverse spectral transform (*IST*) method of Gardner *et al.* (1967). A particular solution is given by

$$\Phi(x,t) = (2q)^{1/2}a \, \frac{\exp \{-4i(v^2 - a^2)t - 2iqv + i\varphi\}}{ch\{2a(x - x_0) + 8avt\}} \tag{7.4}$$

where a characterizes the amplitude, v determines the velocity, x_0 is the centre coordinate, and φ denotes the phase; a and v are independent and they may be chosen arbitrarily.

The N-soliton solution, obtained by superimposition of N one-soliton solutions, will depend therefore on $2N$ arbitrary constants. It will describe, at finite times, the interaction of the N solitons, which will go apart from each other for $t \to \pm \infty$.

This perfunctory presentation of the analytic solution of the 1-dimensional *NLS* equation requires an explanation. The interpretation and/or the modelling of real phenomena will, in most cases, call for a consideration of at least the 2-dimensional and preferably of the 3-dimensional *NLS* equation. Unfortunately, analytical solutions are not available for those equations and therefore recourse will have to be made to their numerical solution. As a first step in that endeavour, the numerical solution of the 1-dimensional *NLS* equation has been studied exhaustively, with a view to the possible extension of the corresponding techniques to the 2- and 3-dimensional cases. It is in this connection that the analytic solution is of significance, as it provides a check for the numerical procedures.

7.3 Numerical Solution of the 1-Dimensional *NLS* Equation

The general approach consists of a separation of the time and space dimensions, with subsequent discretization.

The spatial discretization may be performed within the framework of the *finite difference* or the *finite element* methods (Strang and Fix 1973, Gottlieb and Orszag 1977, Mitchell and Griffiths 1980). A finite difference approach, in which the discrete solutions satisfy the invariance conditions, Eqs. (7.3a) and (7.3b), has been discussed by Delfour *et al.* (1981). But taking into account that it has been found that conservation does not guarantee better results (Sanz Serna and Verwer 1984) and that the finite element method gives the best results (Herbst *et al.* 1985), we will focus our attention on the latter.

7.3.1 The Finite Element Method

Operating within a Hilbert space, multiplication of Eq. (7.1) by ϕ^* (ϕ being a test function) and integration over all the space yields

$$i < \phi | \frac{\partial \Phi}{\partial t} > + < \phi | \frac{\partial^2 \Phi}{\partial x^2} > + q < \phi || \Phi |^2 \Phi > = 0 \qquad (7.5)$$

where the standard quantum-mechanical notation has been used for the expectation values (inner products) for consistency with other chapters. Applying the Green theorem to the second term of Eq. (7.5) yields

$$i < \phi | \frac{\partial \Phi}{\partial t} > - < \frac{\partial \phi}{\partial x} | \frac{\partial \Phi}{\partial x} > + q < \phi || \Phi |^2 \Phi > = 0 \qquad (7.6)$$

which, in a finite-dimensional space with n test functions $\phi_j(x)$, results in the set of equations

$$i < \phi_k | \frac{\partial \Phi}{\partial t} > - < \frac{\partial \phi_k}{\partial x} | \frac{\partial \Phi}{\partial x} > + q < \phi_k || \Phi |^2 \Phi > = 0 \qquad k = 1, 2, ..., n \qquad (7.7)$$

For the practical solution of the problem one will then approximate $\Phi(x,t)$.

$$\Phi(x,t) = \sum_{j=1}^{n} \phi_j(x) w_j(t) \equiv \sum_{j=1}^{n} \phi_j w_j = \phi w \qquad (7.8)$$

as a linear combination of the known test functions ϕ_j, with coefficients w_j to be determined. In this equation, ϕ denotes a row vector with elements ϕ_j and w is a column vector with elements w_j. Within the framework of this approximation one then obtains

$$\frac{\partial \Phi}{\partial t} = \sum_{j=1}^{n} \phi_j(x) \frac{\partial w_j(t)}{\partial t} \equiv \sum_{j=1}^{n} \phi_j w_j' = \phi w' \qquad (7.9)$$

$$\frac{\partial \Phi}{\partial x} = \sum_{j=1}^{n} \frac{\partial \phi_j(x)}{\partial x} w_j(t) \equiv \sum_{j=1}^{n} \phi_j' w_j = \phi' w \qquad (7.10)$$

where ϕ' is a row vector with elements ϕ_j' and w' is a column vector with elements w_j'. [Because of the existence of derivatives with respect to both t and x, differentiation has been denoted with a prime.]

Using the product approximation (Christie *et al.* 1981, Sanz Serna and Christie 1981, Dodd *et al.* 1982, Tourigny 1987) the last term of Eq. (7.7) may be written as

$$<\phi_k \|\Phi\|^2 \Phi> = \sum_{j=1}^{n} <\phi_k|\phi_j>|w_j|^2 w_j \qquad k = 1, 2, ..., n \qquad (7.11)$$

Finally, using Eqs. (7.9)-(7.11), the set of Eqs. (7.7) become

$$\sum_{j=1}^{n} \{ i<\phi_k|\phi_j> w_j' - <\phi_k'|\phi_j'> w_j + q <\phi_k|\phi_j>|w_j|^2 w_j \} = 0 \qquad k = 1, 2, ..., n \qquad (7.12)$$

Using the notation

$$S_{kj} = <\phi_k|\phi_j> \qquad T_{kj} = <\phi_k'|\phi_j'> \qquad U_j = |w_j|^2 w_j$$

we rewrite Eqs. (7.12) as

$$\sum_{j=1}^{n} \{ i S_{kj} w_j' - T_{kj} w_j + q S_{kj} U_j \} = 0 \qquad k = 1, 2, ..., n$$

which, in matrix notation, becomes

$$iSw' - Tw + qSU = 0 \qquad (7.13)$$

where S and T are square symmetric matrices with elements S_{jk} and T_{jk}, respectively, and U is a column vector with elements U_j.

Using the standard piecewise linear functions

$$\phi_j = \begin{cases} (x - x_{j-1})/(x_j - x_{j-1}) & \text{for } x_{j-1} < x \leq x_j \\ (x_{j+1} - x)/(x_{j+1} - x_j) & \text{for } x_j < x \leq x_{j+1} \\ 0 & \text{otherwise} \end{cases}$$

one obtains the expectation values

$$S_{kj} = \frac{h}{6} \qquad T_{kj} = -\frac{1}{h} \qquad \text{for } j = k-1, k+1$$

$$S_{kj} = \frac{2h}{3} \qquad T_{kj} = -\frac{2}{h} \qquad \text{for } j = k \neq 1, n$$

$$S_{kj} = \frac{h}{3} \qquad T_{kj} = \frac{1}{h} \qquad \text{for } j = k = 1, n$$

$$S_{jk} = 0 \qquad T_{jk} = 0 \qquad \text{otherwise}$$

where h is the spacing of the uniform spatial grid.

Multiplication of Eq. (7.13) on the left by iS^{-1} and reordering yields

$$\frac{dw(t)}{dt} = i\{qU - S^{-1}Tw(t)\}$$

which may be written as

$$\frac{dw(t)}{dt} = i\{qW - S^{-1}T\}w(t) = F(w(t))w(t) \tag{7.14}$$

where W is a diagonal matrix with elements $|w_j|^2$ and

$$F(w(t)) = i\{qW - S^{-1}T\}$$

The formal solution of this equation would yield

$$w(t + \tau) = w(t)e^{\int_t^{t+\tau} F(w(\theta))d\theta} \tag{7.15}$$

where τ is the spacing of the uniform time grid.

This approach has been used by Mitchell and Morris (1983), with a time-integration procedure designed (Sanz Serna 1982) in order to conserve the constancy of Eqs. (7.3) while Griffiths *et al.* (1984) use a predictor-corrector iterative scheme for the time integration.

7.3.2 Mesh Refinement Techniques

The solution of the system of non-linear equations using an iterative scheme results in a high cost per time-step. On one hand, large computing times are needed for long-time solutions while on the other hand a large amount of storage is needed when using small discretization values.

These practical difficulties may be overcome by the use of appropriate mesh refinement techniques. The underlying ideas in such techniques are the placing of the spatial grid points where they are needed and to use the largest time-step possible. Diverse approaches may be found in the work of Pereyra and Sewell (1975), White (1979), Russell and Christiansen (1978), Gelinas *et al.* (1981),

Miller and Miller (1981), Davis and Flaherty (1982), Dodd *et al.* (1982), Babuska *et al.* (1983), Manoranjan (1984), Brown and Reyna (1985), Mosher (1985), Thompson (1985), Revilla (1986), Sanz Serna and Christie (1986), Berger and Oliger (1987), and Loh (1987).

The majority of the methods cited above fall into two categories. The first category includes multigrid methods in which a family of grids are superimposed, hierarchically, to refine both in space and in time. These methods typically use *a posteriori* error analysis to identify critical regions in the spatial domain, regions which require finer grids. The other adaptive grid methods are based on the use of equidistribution to generate single non-uniform spatial grids with finer discretization in some areas than others. For example, Manoranjan (1984) suggests the use of arclength equidistribution which generates grids which place more points in regions of high slope than where the solution is flat. Equidistribution methods typically decouple the choice of spatial grids from the time stepping algorithm.

A third approach was proposed by Fraga and Morris (1992). This method aims to produce, as in the equidistribution case, a single non-uniform grid but one which is composed of contiguous piecewise uniform subgrids. The underlying principle is to explicitly place subgrids with fine spacing over regions which exhibit high slopes. These regions are identified using geometry to analyze the current solution profile. Solitons, or soliton-like waves, are located and a grid is generated which places points using a fine uniform subgrid over the regions identified. A much coarser uniform subgrid is used for each interval between pairs of solitons and the intervals between the solitons and the boundaries. The result is a piecewise uniform grid which has been shown to be effective, especially for observing the conservation criteria of the underlying equations. The time stepping algorithm used is, as in the equidistribution methods, independent of the spatial grid method.

More recently, the method has been modified to eliminate large changes in grid spacing over adjacent points in the grid, a problem which is well known to affect non-uniform grids (Russell and Christiansen 1978). Instead of using just one uniform subgrid for the regions between solitons, the non-soliton regions in the spatial domain are discretized using several uniform subgrids, with finer grid spacing near the soliton intervals and progressively coarser grids away from the solitons. This method has been shown to be more robust and also insensitive to the user parameters (Fraga and Morris 1996).

7.4 The Generalized 1-Dimensional *NLS* Equations

A generalized 1-dimensional *NLS*, hereafter denoted as *GNLS*, may be written as

$$i\Phi_t = -\Phi_{xx} + \mathbf{U}(x)\Phi + \mathbf{V}(x,t)\Phi + \mathbf{W}\Phi \qquad (7.16)$$

where $\mathbf{W}$ is a non-linear potential and $\mathbf{U}(x)$ and $\mathbf{V}(x,t)$ are arbitrary potentials.

A specific example is the equation (Johnson 1977, Kakutani and Michihiro 1983)

$$i\Phi_t + \Phi_{xx} + q_c|\Phi|^2\Phi + q_q|\Phi|^4\Phi + i\,q_m|\Phi|^2_x\Phi + i\,q_u|\Phi|^2\Phi_x = 0 \qquad (7.17)$$

which constitutes a model for evolutionary systems with time and space scales greater than those associated with the cubic *NLS* equation (Calogero and Eckhaus 1987). Particular cases of this equation (including the cubic *NLS* equation, discussed in the preceding sections) appear in the study of diverse physical phenomena, such as fluid dynamics (Hasimoto and Ohno 1972), non-linear optics (Strauss 1978), plasma physics (Kaup and Nevell 1978, Lamb 1980), self-modulation of the complex amplitude of solutions of the Benjamin-Ohno equation (Benjamin 1967, Ohno 1975, Tanaka 1982), the propagation of laser beams in an inhomogeneous medium (Pushkarov *et al.* 1979, Cowan *et al.* 1986), and boson gases (Barashenkov and Makhankov 1988). Some of these equations are completely integrable (Kundu 1987, Calogero and Eckhaus 1987) and may be solved by the Inverse Scattering Transform (Calogero and Degasperis 1982) for certain initial conditions.

Equations (7.16) and (7.17) have been studied by Morgan *et al.* (1997) and Pathria and Morris (1989, 1990), respectively, and their formulations are examined below.

7.4.1 Equation $i\Phi_t + \Phi_{xx} + q_c|\Phi|^2\Phi + q_q|\Phi|^4\Phi + i\,q_m|\Phi|^2_x\Phi + i\,q_u|\Phi|^2\Phi_x = 0$

Pathria and Morris (1989) use the gauge transformation

$$\Phi \equiv \Phi(x,t) = \psi(x,t)e^{i\theta(x,t)} \equiv \psi\,e^{i\theta} \qquad (7.18)$$

which is a generalization of that used by Calogero and Eckhaus (1987a) and Kundu (1987), in order to obtain a new equation from which the exact solutions of Eq. (7.17) may then be found. ψ is a complex-valued function and θ is a real function defined by

$$\theta_x = 2k|\psi|^2 \qquad (7.19a)$$

$$\theta_t = 4k\,\mathrm{Im}(\psi\psi_x^*) \qquad (7.19b)$$

where *Im* $(\psi\psi_x)$ stands for the imaginary part of $(\psi\psi_x)$.

Using the above transformation, with $k = -(2q_m + q_u)/8$, in Eq. (7.17) yields

$$i\,\psi_t + \psi_{xx} + q_c|\psi|^2\psi + [q_q - \frac{1}{16}(2q_m+q_u)(2q_m-3q_u)]|\psi|^4\psi$$

$$+ \frac{i}{2}(q_m - \frac{1}{2}q_u)|\psi|_x^2\psi + i(-q_m + \frac{1}{2}q_u)|\psi|^2\psi_x + \frac{1}{2}(2q_m+q_u)\mathrm{Im}\,(\psi\psi_x^*)\psi = 0$$

which may be rewritten, after some manipulation of the last three terms, as

$$i\psi_t + \psi_{xx} + F(x)\psi = 0 \tag{7.20}$$

where

$$F(x) = Q_c|\psi|^2 + Q_q|\psi|^4 + Q_u\mathrm{Im}(\psi\psi_x^*) \tag{7.21}$$

is a real function and

$$Q_c = q_c \qquad Q_q = q_q - \frac{1}{16}(2q_m+q_u)(2q_m-3q_u) \qquad Q_u = q_u$$

The functions $\theta(x,t)$ and $\psi(x,t)$ may now be obtained using

$$\psi \equiv \psi(x,t) = f(x\text{-}ct)\,e^{ig(x\text{-}bt)} \equiv f\,e^{ig} \tag{7.22}$$

where f and g are real functions, with

$$g(x\text{–}bt) = \frac{c}{2}(x\text{–}bt) + v$$

and c, b, and v are arbitrary constants. The function $\theta(x,t)$ is obtained immediately as

$$\theta(x,t) = 2k\int \psi^*\,\psi\,dx = 2k\int f^2(x\text{–}ct)\,d(x\text{–}ct) \tag{7.23}$$

by substitution of Eq. (7.22) into Eq. (7.19a) and integration over x or, equivalently, over (x-ct). On the other hand, use of Eq. (7.22) into Eq. (7.20) yields

$$i\{f_t + 2g_x f_x + g_{xx}f\} + \{-g_t f + f_{xx} - g_x^2 f + Q_c f^3 + Q_q f^5 - Q_u g_x f^3\} = 0$$

which reduces to

$$f'' + Q_q f^5 + (Q_c - \frac{c}{2}Q_u)f^3 + \frac{c}{2}(b - \frac{c}{2})f = 0$$

where the primes denote differentiation with respect to (x-ct). Multiplying this equation by f' and integrating up once yields

$$(f')^2 + \frac{1}{3} Q_q f^6 + \frac{1}{2} (Q_c - \frac{c}{2} Q_u) f^4 + \frac{c}{2} (b - \frac{c}{2}) f^2 + C = 0 \qquad (7.24)$$

where C is an integration constant. With the substitution $z = f^2$, Eq. (7.24) transforms into

$$(z')^2 + \frac{4}{3} Q_q z^4 + (2Q_c - cQ_u)z^3 + (2bc-c^2)z^2 + 4Cz = 0 \qquad (7.25)$$

which may be solved in terms of elliptic functions. In the case when $C = 0$, the solutions may be expressed in terms of elementary functions and include oscillatory, singular, phase jump, and solitary wave solutions. The parameters *b* and *c*, which represent the speeds of the carrier and envelope waves of ψ, determine in part the form of the solution.

Additional solutions may then be obtained by means of a Galileo transformation (see, e.g., Born 1962 and Stephenson and Kilmister 1987). The transformation

$$X = A^2(x + 2A^2Bt) \qquad (7.26a)$$

$$\tau = A^4 t \qquad (7.26b)$$

$$\phi = \frac{1}{A} \psi e^{iA^2B(x + A^2Bt)} \qquad (7.26c)$$

where

$$B = \frac{q_c}{q_u} (1 - \frac{1}{A^2})$$

and *A* is an arbitrary, non-zero real constant, yields (with $q_u \neq 0$)

$$i\phi_\tau + \phi_{XX} + q_c|\phi|^2\phi + q_q|\phi|^4\phi + iq_m|\phi|^2_X\phi + iq_u|\phi|^2\phi_X = 0$$

which shows the invariance of Eq. (7.17) under the transformation. Therefore, given a function ψ, new solutions may be obtained by means of Eq. (7.26c).

Pathria and Morris (1989) have derived, from Eq. (7.20) and assuming that ψ is a rapidly decreasing function of *x* as $x \rightarrow \pm \infty$, the conservation laws

$$N = \int_{-\infty}^{\infty} |\psi|^2 \, dx$$

$$H = \int_{-\infty}^{\infty} [4 \operatorname{Im}(\psi \psi_x^*) + Q_u |\psi|^4] dx$$

$$P = \int_{-\infty}^{\infty} [|\psi|^2 - \frac{1}{2} Q_c |\psi|^4 - \frac{1}{3} Q_q |\psi|^6] dx$$

The reader is referred to the work of Pathria and Morris (1989) for the proofs of these laws as well as for the conditions for boundedness and blow-up of the solutions of Eqs. (7.17) and (7.20). Numerical methods based on collocation are described by Pathria and Morris (1990, 1991), Robinson and Fairweather (1994) and Robinson (1997).

7.4.2 Equation $i\Phi_t = -\Phi_{xx} + U(x)\Phi + V(x,t)\Phi + W\Phi$

In the case when $W \propto |\Phi|^2$ [see Eq. (7.1)], Eq. (7.16) is denoted as the Gross-Pitaevskii equation (Ginzburg and Pitaevskii 1958, Gross 1963), used in the study of the Bose-Einstein condensation in inhomogeneous dilute alkali gases (Anderson et al. 1995, Davis et al. 1995). The Gross-Pitaevskii equation has been studied by Ruprecht et al. (1995), Bayn and Pethick (1996), Edwards et al. (1996), and Holland and Cooper (1996).

The general case, Eq. (7.16), with arbitrary potentials, has been treated by Morgan et al. (1997), with possible generalization to a higher number of dimensions (see below). The corresponding formulation is summarized below but the reader is referred to the work of Morgan et al. (1997) for the details of the physical implications.

As in earlier treatments of the one-dimensional case (Chen and Liu 1976, 1978, Hasse 1982, Nassar 1986, de Moura 1988, 1994, Nogami and Toyama 1994ab; see also Hussimi 1953 and Kerner 1958) the function is decomposed into a modulus (depending on time only via a change in its mean position) and a phase factor. The solution is written as

$$\Phi_{new}(x,t) = \Phi_{old}(x-x_0(t),t) e^{iS(x,t)} \tag{7.27}$$

with

$$\Phi_{old}(x-x_0(t),t) = \phi(x-x_0(t)) e^{-i\varepsilon t}$$

and where $\phi(x)$ is the solution to the fixed-potential problem

$$\varepsilon\phi(x) = -\frac{d^2\phi(x)}{dx^2} + \mathcal{U}(x)\phi(x) + \mathcal{W}\phi(x) \tag{7.28}$$

The energy is denoted by ε and $x_0(t)$ is the time-dependent shift in the position of the function, which may be expressed as

$$x_0(t) = \mu(t) - \mu_0 \equiv \mu(t) - \mu_0$$

so that, in general,

$$\mu(t) = <x> = \int x|\phi(x)|^2 dx$$

$S(x,t)$ must be real in order that the propagation of the original solution will take place without change of shape. Morgan *et al.* (1997) emphasize that the assumption about the existence of solutions to the fixed-potential problem will be valid only if the combination of fixed potential and non-linearity allows such solutions.

Substitution of $\Phi_{new}(x,t)$, Eq. (7.27), into Eq. (7.16), with change of variables from (x,t) to (q,t), where $q = x - x_0(t)$ is the internal coordinate of the solitary wave, yields

$$i\frac{\partial\phi(q)}{\partial t} - \phi(q)\frac{\partial S(q,t)}{\partial t} = -i\{2\frac{\partial\phi(q)}{\partial q}\frac{\partial S(q,t)}{\partial q} + \phi(q)\frac{\partial^2 S(q,t)}{\partial q^2}\}$$
$$+ \phi(q)\left(\frac{\partial S(q,t)}{\partial q}\right)^2 + \Delta\mathcal{U}(q + x_0(t), x_0(t))\,\phi(q) + \mathcal{V}(q + x_0(t), t)\,\phi(q) \tag{7.29}$$

after elimination of the exponential factors and taking into account Eq. (7.28), and where

$$\Delta\mathcal{U}(q + x_0(t), x_0(t)) = \mathcal{U}(x) - \mathcal{U}(q)$$

with the added assumption that the non-linear potential depends only on the internal coordinate.

Using

$$\phi(q) = \rho(q)\,e^{i\Omega(q)}$$

where $\rho(q)$ and $\Omega(q)$ are real functions, Eq. (7.29) may be separated into the real and imaginary parts

$$-\frac{\partial\Omega(q)}{\partial t} - \frac{\partial S(q,t)}{\partial t} = 2\frac{\partial S(q,t)}{\partial q}\frac{d\Omega(q)}{dq} + \left(\frac{\partial S(q,t)}{\partial q}\right)^2$$
$$+ \Delta \mathcal{U}(q + x_0(t), x_0(t)) + \mathcal{V}(q + x_0(t), t) \tag{7.30a}$$

$$\frac{\partial\rho(q)}{\partial t} = -2\frac{\partial\rho(q)}{\partial q}\frac{\partial S(q,t)}{\partial q} - \rho(q)\frac{\partial^2 S(q,t)}{\partial q^2} \tag{7.30b}$$

respectively, after having eliminated the exponential factor in both equations and having divided by ρ ($\neq 0$) in the first one. After multiplication by $\rho(q)$ and using

$$\frac{\partial\rho(q)}{\partial t} = \frac{d\rho(q)}{dq}\frac{dq}{dt} = -\frac{d\rho(q)}{dq}\frac{dx_0(t)}{dt}$$

Eq. (7.30b) transforms into

$$\rho^2(q)\frac{\partial^2 S(q,t)}{\partial q^2} + 2\rho(q)\frac{d\rho(q)}{dq}\frac{\partial S(q,t)}{\partial q} = \frac{d}{dq}\{\rho^2(q)\frac{\partial S(q,t)}{\partial q}\}$$
$$= \rho(q)\frac{d\rho(q)}{dq}\frac{dx_0(t)}{dt} = \frac{d}{dq}\{\frac{1}{2}\rho^2(q)\}\frac{dx_0(t)}{dt}$$

which yields, after integration over q,

$$\rho^2(q)\frac{\partial S(q,t)}{\partial q} = \frac{1}{2}\rho^2(q)\frac{dx_0(t)}{dt} + C_1(t)$$

In the particular case when $C_1(t) = 0$, appropriate for localized solutions as well as when $\mathcal{U}(x) = 0$ and $\mathcal{V}(x,t) = 0$ and $\mathcal{W} \propto |\Phi|^2$, one obtains

$$\frac{\partial S(q,t)}{\partial q} = \frac{\partial S(x,t)}{\partial x} = \frac{1}{2}\frac{dx_0(t)}{dt} \tag{7.31}$$

which, upon new integration, yields

$$S(x,t) = \frac{1}{2}\frac{dx_0(t)}{dt}x + C_2(t) \tag{7.32}$$

On the other hand, using Eq. (7.31) to express

$$\left(\frac{\partial S(q,t)}{\partial q}\right)^2 = \frac{1}{4}\left(\frac{dx_0(t)}{dt}\right)^2$$

$$\frac{\partial\Omega(q)}{\partial t} + 2\frac{\partial S(q,t)}{\partial q}\frac{\partial\Omega(q)}{\partial q} = \frac{\partial\Omega(q)}{\partial q}\frac{\partial q}{\partial t} + 2\frac{\partial S(q,t)}{\partial q}\frac{\partial\Omega(q)}{\partial q}$$

$$= -\frac{\partial\Omega(q)}{\partial q}\frac{dx_0(t)}{dt} + 2\left(\frac{1}{2}\frac{dx_0(t)}{dt}\right)\frac{\partial\Omega(q)}{\partial q} = 0$$

and reverting to the x-coordinate description, Eq. (7.30a) becomes

$$-\frac{\partial S(x,t)}{\partial t} = \frac{1}{4}\left(\frac{dx_0(t)}{dt}\right)^2 + \Delta\mathcal{U}(x,x_0(t)) + \mathcal{V}(x,t) \tag{7.33}$$

which, upon integration over t, yields

$$S(x,t) = -\frac{1}{4}\int\left(\frac{dx_0(t)}{dt}\right)^2 dt - \int\{\Delta\mathcal{U}(x,x_0(t)) + \mathcal{V}(x,t)\}dt + C_3(x) \tag{7.34}$$

Equations (7.32) and (7.34) must be consistent with each other in order that solitary wave motion will be possible. The corresponding condition is obtained by differentiating Eq. (7.31) with respect to t and Eq. (7.33) with respect to x and equating the resulting equations. One obtains, taking into account that $x_0(t)$ depends only on t, the equation of motion

$$\frac{1}{2}\frac{d^2x_0(t)}{dt^2} = \frac{1}{2}\frac{d^2\mu(t)}{dt^2} = -\frac{\partial}{\partial x}\{\Delta\mathcal{U}(x,x_0(t)) + \mathcal{V}(x,t)\} \tag{7.35}$$

The *lhs* of this equation depends only on t and therefore the *rhs* must also depend only on t for propagation without change of shape. That is, the condition for solitary wave motion to exist is that the quantity within the curly brackets must be at most a linear function of x.

7.5 The Problem of Higher Dimensions

As pointed out in Section 7.2.2, the modelling of real phenomena will require the solution of higher-order *NLS* equations. Although the formulation described above may be extended, under specific conditions, to higher dimensions, in general it may be necessary to resort to numerical methods.

The numerical solution of the *NLS* equation in two or three dimensions requires that the following points be addressed: the selection of an appropriate grid, the interpolation of values to this grid, the efficient description of a non-uniform grid, and the use of a numerical scheme appropriate for both the grid and the partial differential equation under study (Fraga 1988).

The modified algorithm (Fraga and Morris (1996) mentioned above may, in fact, be of use in this connection insofar as piecewise uniformity, gradual change in spacing, and the geometric approach to the determination of the areas of refinement are concerned.

This algorithm has been used in the study of the radial *NLS* equation

$$i\frac{\partial\Phi}{\partial t} + \frac{\partial^2\Phi}{\partial\rho^2} + \frac{1}{\rho}\frac{\partial\Phi}{\partial\rho} + |\Phi^2|\Phi = 0$$

where again Φ is a complex-value function, $t \geq 0$, and $\rho = (x^2 + y^2)^{1/2}$. The paths of the soliton-like objects in the solutions were predicted (Tourigny *et al.* 1988) to be hyperbolic and long-time solutions were necessary in order to verify this behaviour. The results show that, depending on the direction of motion, the amplitude decreases/increases and the support widens/narrows, indicating that the algorithm caters dynamically to the soliton-like object even though its support is changing.

The Riccati Equation

The Riccati equation first arose, as a mathematical curiosity, from the dabblings of an Italian nobleman during the Renaissance and was soon forgotten except by the members of the Bernoulli family, intent on its solution.

In any case, sooner or later, it would have appeared either within the context of differential calculus *per se* or in the analysis of new problems in modern physics and in the formulation of new theories (such as soliton theory). In fact, the Riccati equation has been associated with quantum mechanics since the early days of the latter, although that connection has been largely ignored until recently.

The Riccati equation has been studied in detail by now. Although it cannot be solved by elementary methods, its general solution may be obtained in some cases, as discussed in Chapter 8. Chapter 9 will be dedicated mostly to the use of the Riccati equation in connection with the 1-dimensional linear Schrödinger equation and the related Sturm-Liouville problem.

8 The Riccati Equation and Its Solution

8.1 Derivation of the Riccati Equation

We could proceed directly with an inspection of the solution of the Riccati equation but we believe that it is more appropriate to establish first its connection with other differential equations. In this connection, as well as for further details, the reader is referred to the work of Simmons (1972), Gradshteyn and Ryzhik (1980), Bronshtein and Semendyayev (1985), and Gellert *et al.* (1989).

A second-order, homogeneous linear differential equation

$$U(\zeta) \frac{d^2\eta}{d\zeta^2} + V(\zeta) \frac{d\eta}{d\zeta} + W(\zeta)\eta = 0 \qquad (8.1)$$

may be made self-adjoint by multiplication with $S(\zeta)/U(\zeta)$, where

$$S(\zeta) = \exp\left\{ \int \frac{V(\zeta)}{U(\zeta)} d\zeta \right\}$$

The resulting equation

$$S(\zeta) \frac{d^2\eta}{d\zeta^2} + \frac{V(\zeta)}{U(\zeta)} S(\zeta) \frac{d\eta}{d\zeta} + \frac{W(\zeta)}{U(\zeta)} S(\zeta)\eta = 0 \qquad (8.2a)$$

may be written as

$$\frac{d}{d\zeta}[S(\zeta)\frac{d\eta}{d\zeta}] + T(\zeta)\eta = 0 \qquad (8.2b)$$

where

$$T(\zeta) = \frac{W(\zeta)}{U(\zeta)} S(\zeta)$$

$S(\zeta)$ and $T(\xi)$ are continuous in the interval [a,b] and $S(\zeta) > 0$. [The notation used has been adopted in order to be consistent with the formulation presented in Chapters 10 and 11 and for simplicity in the discussion of the solution of the Riccati equation, presented below.]

Using

$$U(\zeta) = \frac{1}{R(\zeta)}$$

$$V(\zeta) = -\frac{Q(\zeta)}{R(\zeta)} - \frac{1}{R^2(\zeta)} \frac{dR(\zeta)}{d\zeta}$$

$$W(\zeta) = P(\zeta)$$

and eliminating the factor $S(\zeta)$, Eq. (8.2) becomes

$$\frac{d^2\eta}{d\zeta^2} - R(\zeta) \left\{ \frac{Q(\zeta)}{R(\zeta)} + \frac{1}{R^2(\zeta)} \frac{dR(\zeta)}{d\zeta} \right\} \frac{d\eta}{d\zeta} + P(\zeta)R(\zeta)\eta = 0$$

which may be rewritten as

$$\frac{1}{\eta R(\zeta)} \frac{d^2\eta}{d\zeta^2} - \left\{ \frac{Q(\zeta)}{\eta R(\zeta)} + \frac{1}{\eta R^2(\zeta)} \frac{dR(\zeta)}{d\zeta} \right\} \frac{d\eta}{d\zeta} + P(\zeta) = 0 \qquad (8.3)$$

With the transformation

$$x = -\frac{1}{\eta R(\zeta)} \frac{d\eta}{d\zeta}$$

one obtains

$$\frac{dx}{d\zeta} = \frac{1}{\eta^2 R(\zeta)} \left(\frac{d\eta}{d\zeta} \right)^2 + \frac{1}{\eta R^2(\zeta)} \frac{dR(\zeta)}{d\zeta} \frac{d\eta}{d\zeta} - \frac{1}{\eta R(\zeta)} \frac{d^2\eta}{d\zeta^2}$$

$$= R(\zeta)x^2 + \frac{1}{\eta R^2(\zeta)} \frac{dR(\zeta)}{d\zeta} \frac{d\eta}{d\zeta} - \frac{1}{\eta R(\zeta)} \frac{d^2\eta}{d\zeta^2}$$

so that

$$\frac{1}{\eta R(\zeta)} \frac{d^2\eta}{d\zeta^2} = -\frac{dx}{d\zeta} + R(\zeta)x^2 + \frac{1}{\eta R^2(\zeta)} \frac{dR(\zeta)}{d\zeta} \frac{d\eta}{d\zeta}$$

Substitution in Eq. (8.3) yields

$$-\frac{dx}{d\zeta} + R(\zeta)x^2 + \frac{1}{\eta R^2(\zeta)}\frac{dR(\zeta)}{d\zeta}\frac{d\eta}{d\zeta} - \left\{\frac{Q(\zeta)}{\eta R(\zeta)} + \frac{1}{\eta R^2(\zeta)}\frac{dR(\zeta)}{d\zeta}\right\}\frac{d\eta}{d\zeta} + P(\zeta)$$

$$= -\frac{dx}{d\zeta} + R(\zeta)x^2 - \frac{Q(\zeta)}{\eta R(\zeta)}\frac{d\eta}{d\zeta} + P(\zeta)$$

$$= -\frac{dx}{d\zeta} + R(\zeta)x^2 + Q(\zeta)x + P(\zeta) = 0$$

which will be rewritten as

$$\frac{dx}{d\zeta} = P(\zeta) + Q(\zeta)x + R(\zeta)x^2 \tag{8.4}$$

This is the *general* Riccati equation, which constitutes an extension of the simple first-order, linear differential equation.
    A further substitution

$$x = X \exp\left\{\int_a^\zeta Q(\zeta)d\zeta\right\} = XL(a,\zeta)$$

yields

$$\frac{dx}{d\zeta} = \left\{\frac{dX}{d\zeta} + Q(\zeta)X\right\}L(a,\zeta) = P(\zeta) + Q(\zeta)XL(a,\zeta) + R(\zeta)[XL(a,\zeta)]^2$$

which becomes

$$L(a,\zeta)\frac{dX}{d\zeta} = P(\zeta) + R(\zeta)[XL(a,\zeta)]^2$$

In the particular case when $a = 0$, we obtain the *transformed* Riccati equation

$$\frac{dX}{d\zeta} = M(\zeta) + N(\zeta)X^2 \tag{8.5}$$

where

$$M(\zeta) = P(\zeta) \exp\left\{-\int_0^\zeta Q(\zeta)d\zeta\right\} \tag{8.6a}$$

$$N(\zeta) = R(\zeta) \exp \left\{ \int_0^\zeta Q(\zeta)d\zeta \right\}$$ (8.6b)

Finally, the *special* Riccati equation results from the general equation when

$$P(\zeta) = a\,\zeta^m \qquad\qquad Q(\zeta) = 0 \qquad\qquad R(\zeta) = b$$ (8.7)

where a and b are constants.

8.2 Solution of the Riccati Equation

The Riccati equation cannot be solved by elementary methods, but its general solution may be obtained in certain instances.

8.2.1 Solution of the General Riccati Equation

If a particular solution (x_1) is known, the general solution may then be written as

$$x = x_1 + y$$

so that substitution into Eq. (8.4) yields

$$\frac{dx_1}{d\zeta} + \frac{dy}{d\zeta} = P(\zeta) + Q(\zeta)(x_1 + y) + R(\zeta)(x_1^2 + 2x_1y + y^2)$$

Taking into account that x_1 is a solution of Eq. (8.4), the preceding equation reduces to

$$\frac{dy}{d\zeta} = [Q(\zeta) + 2R(\zeta)x_1]y + R(\zeta)y^2$$ (8.8)

which is the Bernoulli differential equation. The substitution

$$y = \frac{1}{z} \qquad\qquad \frac{dy}{d\zeta} = -\frac{1}{z^2}\frac{dz}{d\zeta}$$

yields in turn

$$\frac{dz}{d\zeta} + [Q(\zeta) + 2R(\zeta)x_1]z = -R(\zeta)$$

or, in short,

$$\frac{dz}{d\zeta} + p_1(\zeta)z = q(\zeta)$$ (8.9)

with

$$p_1(\zeta) = Q(\zeta) + 2R(\zeta)x_1 \qquad q(\zeta) = -R(\zeta)$$ (8.10)

Equation (8.9) is a first-order, linear differential equation, whose general solution is of the form

$$z = \frac{v_1(\zeta) + c_{(1)}}{u_1(\zeta)}$$ (8.11)

where $c_{(1)}$ is an integration constant and

$$u_1(\zeta) = \exp\left\{\int p_1(\zeta)d\zeta\right\} \qquad v_1(\zeta) = \int q(\zeta)u_1(\zeta)d\zeta$$ (8.12)

Therefore the general solution of the Riccati equation will be

$$x = x_1 + \frac{u_1(\zeta)}{v_1(\zeta) + c_{(1)}}$$ (8.13)

Let us now assume that several particular solutions of the Riccati equation are known. For the i-th solution, x_i, Eqs. (8.11) and (8.13) allow us to write, in general,

$$z_{(i)} = \frac{1}{x - x_i} = \frac{v_i(\zeta) + c_{(i)}}{u_i(\zeta)}$$

while for any two solutions, x_i and x_j, we can write

$$z_{j(i)} = \frac{1}{x_j - x_i} = \frac{v_i(\zeta) + c_{j(i)}}{u_i(\zeta)}$$

Therefore, in the case when three particular solutions (x_1, x_2, x_3) are known we will be able to write

$$\frac{x - x_2}{x - x_1} \bigg/ \frac{x_3 - x_2}{x_3 - x_1} = \frac{[v_1(\zeta) + c_{(1)}] [v_2(\zeta) + c_{3(2)}]}{[v_1(\zeta) + c_{3(1)}] [v_2(\zeta) + c_{(2)}]}$$

or, in short,

$$\frac{x - x_2}{x - x_1} \Big/ \frac{x_3 - x_2}{x_3 - x_1} = C$$

At the numerical level, an iterative approach has been used successfully (Lebeda and Thorson 1970, Newman and Thorson 1972). Rewriting Eq. (8.4) as

$$x = \alpha_0(\zeta) + \beta_0(\zeta)x^2 + \gamma_0(\zeta)x'$$

where the prime denotes derivative, and using the substitution $x = \alpha_0(\zeta) + x_1$, it is found that x_1 satisfies

$$x_1 = \alpha_1(\zeta) + \beta_1(\zeta)x_1^2 + \gamma_1(\zeta)x_1'$$

a Riccati equation of the same form as the original one. Repeating the process one arrives at

$$x = \sum_{j=0}^{n} \alpha_j(\zeta) + x_{n+1}$$

$$x_{n+1} = \alpha_{n+1}(\zeta) + \beta_{n+1}(\zeta)x_{n+1}^2 + \gamma_{n+1}(\zeta)x_{n+1}'$$

with

$$\alpha_{n+1}(\zeta) = \frac{\alpha_n^2(\zeta)\beta_n(\zeta) + \alpha_n'(\zeta)\gamma_n(\zeta)}{1 - 2\alpha_n(\zeta)\beta_n(\zeta)}$$

$$\beta_{n+1}(\zeta) = \frac{\beta_n(\zeta)}{1 - 2\alpha_n(\zeta)\beta_n(\zeta)}$$

$$\gamma_{n+1}(\zeta) = \frac{\gamma_n(\zeta)}{1 - 2\alpha_n(\zeta)\beta_n(\zeta)}$$

It should be mentioned, however, that the actual convergence properties of the procedure are not clear and that they will depend on the properties of $\alpha_0(\zeta)$, $\beta_0(\zeta)$, and $\gamma_0(\zeta)$ (Newman and Thorson 1972).

8.2.2 Solution of the Transformed Riccati Equation

In this case we could also apply the procedures outlined above, with the appropriate modification due to the absence of the term in X in Eq. (8.4). Alternately one may proceed as follows.

Let us consider the second-order, homogeneous linear differential equation

$$\frac{d}{d\zeta}\left\{-\frac{1}{N(\zeta)}\frac{d\xi}{d\zeta}\right\} - M(\zeta)\xi = -\frac{1}{N(\zeta)}\xi'' + \frac{1}{N^2(\zeta)}\frac{dN(\zeta)}{d\zeta}\xi' - M(\zeta)\xi = 0 \qquad (8.14)$$

where the abbreviated standard notation for the derivatives has been used, in order to simplify the following developments, and $M(\zeta)$ and $N(\zeta)$ are the quantities appearing in Eq. (8.5). If ξ_1 and ξ_2 are two particular solutions of the above equation, substitution of

$$X = -\frac{1}{N(\zeta)}\frac{c_1\xi_1' + c_2\xi_2'}{c_1\xi_1 + c_2\xi_2} \qquad (8.15)$$

into Eq. (8.5) yields

$$\frac{dX}{d\zeta} - M(\zeta) - N(\zeta)X^2$$

$$= \frac{1}{N^2(\zeta)}\frac{dN(\zeta)}{d\zeta}\frac{c_1\xi_1' + c_2\xi_2'}{c_1\xi_1 + c_2\xi_2} - \frac{1}{N(\zeta)}\frac{c_1\xi_1'' + c_2\xi_2''}{c_1\xi_1 + c_2\xi_2} + \frac{1}{N(\zeta)}\frac{(c_1\xi_1' + c_2\xi_2')^2}{(c_1\xi_1 + c_2\xi_2)^2}$$

$$- M(\zeta) - N(\zeta)\frac{1}{N^2(\zeta)}\frac{(c_1\xi_1' + c_2\xi_2')^2}{(c_1\xi_1 + c_2\xi_2)^2}$$

$$= \frac{1}{c_1\xi_1 + c_2\xi_2}\left\{-\frac{1}{N(\zeta)}(c_1\xi_1'' + c_2\xi_2'') + \frac{1}{N^2(\zeta)}\frac{dN(\zeta)}{d\zeta}(c_1\xi_1' + c_2\xi_2')\right.$$

$$\left. - M(\zeta)(c_1\xi_1 + c_2\xi_2)\right\}$$

Taking into account Eq. (8.14) one can say that this expression will vanish. That is, X, defined by Eq. (8.15), will be a solution of the transformed Riccati equation.

In this connection it must be mentioned (Simmons 1972) that $(c_1\xi_1 + c_2\xi_2)$ represents the general solution of the second-order differential equation above and that the second solution ξ_2 may be obtained from ξ_1 as

$$\xi_2 = \xi_1\int\left\{\frac{1}{\xi_1^2}\exp\left[\frac{1}{N^2(\zeta)}\frac{dN(\zeta)}{d\zeta}\right]d\zeta\right\} \qquad (8.16)$$

In the particular case when the transformed Riccati equation has been obtained from the general Riccati equation, taking into account Eq. (8.6b), one obtains

$$\frac{dN(\zeta)}{d\zeta} = \frac{dR(\zeta)}{d\zeta} \exp\left\{\int_0^\zeta Q(\zeta)d\zeta\right\} + R(\zeta)Q(\zeta) \exp\left\{\int_0^\zeta Q(\zeta)d\zeta\right\} \tag{8.17a}$$

$$\frac{1}{N^2(\zeta)}\frac{dN(\zeta)}{d\zeta} = \left\{\frac{1}{R^2(\zeta)}\frac{dR(\zeta)}{d\zeta} + \frac{Q(\zeta)}{R\zeta}\right\} \exp\left\{-\int_0^\zeta Q(\zeta)d\zeta\right\} \tag{8.17b}$$

so that, if x_1 is a particular solution of the general Riccati equation, then

$$X_1 = x_1 \exp\left\{-\int Q(\zeta)d\zeta\right\} \tag{8.18}$$

is a particular solution of the transformed Riccati equation.

8.2.3 Solution of the Special Riccati Equation

The special Riccati equation is soluble when m = 0 or m = 4k/(1-2k), with $k = \pm 1$, $\pm 2, \dots$. As this equation is of no interest for the purpose of this work, the reader is referred to the work of Bronshtein and Semendyayev (1985) for further details.

9 Quantum-Mechanical Applications of the Riccati Equation

9.1 Introduction

The Riccati equation has found important applications in quantum-mechanical problems, such as, for example, high-order perturbation calculations (Aharonov and Au 1979, Eletsky and Popov 1980, Privman 1981). In this work, however, we will discuss its application to the solution of the Sturm-Liouville problem and of the 1-dimensional Schrödinger equation, which represents a particular case of the Sturm-Liouville equation.

Consequently it is convenient to start this chapter with a brief examination of the Sturm-Liouville equation and of the development that leads to a Riccati equation.

The notation adopted in Chapter 8, as noted there, was chosen with consideration of the developments to be presented in that chapter as well as in Chapters 10 and 11. In this chapter it will be modified in order to better illustrate the connection between the Sturm-Liouville and the 1-dimensional Schrödinger equations.

9.1.1 The Sturm-Liouville Equation

The Sturm-Liouville equation is a self-adjoint, second-order linear differential equation. Therefore it may be expressed in general by Eq. (8.2), that is,

$$\frac{d}{dx}[S(x)\frac{d\Psi(\varepsilon,x)}{dx}] + T(\varepsilon,x)\Psi(\varepsilon,x) = 0 \tag{9.1}$$

where customarily $T(\varepsilon,x)$ is expressed as

$$T(\varepsilon,x) = -V(x) + \varepsilon W(x) \tag{9.2}$$

In this connection it should be mentioned that no relationship is intended between the present quantities V(x) and W(x) and those used in Chapter 8. In particular, the notation V(x) has been adopted here because it will be seen to correspond to the potential in the 1-dimensional Schrödinger equation. Hereafter, for simplicity, the function will be denoted by either $\Psi(\varepsilon,x)$ or $\Psi(x)$, depending on whether its ε-dependence is to be emphasized or not.

The conditions holding for the quantities S(x), V(x), and W(x) are: S(x), V(x), and W(x) are continuous in the open interval (a,b); S(x) and V(x) are differentiable in (a,b); S(x) is assumed to be positive (in order to avoid the existence of singular points in the interval); W(x) does not change sign in the closed interval [a,b]. W(x) and ε have been denoted (Pauling and Wilson 1935) as the weight factor and the characteristic-value parameter, respectively; the latter will represent the eigenvalue in the 1-dimensional Schrödinger equation.

The interval may be finite or infinite and the endpoints may be singular or non-singular. The properties of the solutions in the neighbourhood of the endpoints (for the existence of a discrete spectrum of eigenvalues through a finite/infinite interval of real values of ε) and the completeness of the set of discrete eigenfunctions (independently of whether the discrete spectrum extends or not to $+\infty$) have been studied in detail (Weyl 1910, Fues 1926, Oppenheimer 1927, 1928, Courant and Hilbert 1931, Kemble 1958).

In addition to the 1-dimensional Schrödinger equation, diverse equations of the form of the Sturm-Liouville equation appear in both quantum mechanics and theoretical physics; in particular, the modified Bessel, Hermite, Legendre hypergeometric, Jacobi, Laguerre, Legendre, and Tschebycheff differential equations represent particular cases of the Sturm-Liouville equation (Franklin 1960).

9.1.2 The 1-Dimensional Schrödinger Equation

With $S(x) = \hbar^2/2m$ and $W(x) = 1$, in the interval $0 \leq x \leq \infty$, the Sturm-Liouville equation becomes

$$-\frac{\hbar^2}{2m}\frac{d^2\Psi(x)}{dx^2} + \mathcal{V}(x)\,\Psi(x) = E\Psi(x) \tag{9.3}$$

which is the 1-dimensional Schrödinger equation for a single particle (with mass m) subject to the potential $\mathcal{V}(x)$ and where the parameter ε has been replaced with E in order to conform to standard notation.

In a general way, the same designation is also given to the equation

$$\frac{d^2\Psi(x)}{dx^2} = [V(x) - E]\Psi \tag{9.4}$$

which appears in molecular spectroscopy and theoretical physics.

Considerable effort has been devoted to the development of efficient algorithms for the numerical solution of the 1-dimensional Schrödinger equation. As here we will restrict our attention to those formulations involving the Riccati equation, we will refer the reader to the work of Blatt (1967), Newman and Thorson (1972), Raptis and Allison (1978), Ixaru (1980, 1984), Ixaru and Rizea (1980), Raptis (1982), Cash and Raptis (1984), Raptis and Cash (1985, 1987), Liu and Dykstra (1986), Dykstra (1987), Dykstra et al. (1987), Papageorgiou and Raptis (1987), Cash et al. (1990), Papageorgiou et al. (1990), Simos (1990, 1991, 1992), Searles and von Nagy-Felsobuki (1992), Wang et al. (1992), Franken and Dykstra (1993), Parish and Dykstra (1993), Simos and Tougelidis (1996), and Gonzales et al. (1997).

9.1.3 The Riccati Equation for the Logarithmic Derivative

For simplicity, hereafter we will be concerned exclusively with Sturm-Liouville equations of the form

$$\frac{d^2\Psi(x)}{dx^2} + A(x)\frac{d\Psi(x)}{dx} + B(E,x)\Psi(x) = 0 \qquad (9.5)$$

which includes Eq. (9.3) and (9.4) as particular cases.
    Using the transformation

$$\phi(x) = \ell n \Psi(x)$$

(hence the designation *logarithmic*) we obtain

$$\Psi(x) = e^{\phi(x)}$$

$$\frac{d\Psi(x)}{dx} = \frac{d\phi(x)}{dx} e^{\phi(x)} = \psi(x)e^{\phi(x)}$$

$$\frac{d^2\Psi(x)}{dx^2} = \{\frac{d\psi(x)}{dx} + \psi^2(x)\} e^{\phi(x)}$$

with

$$\psi(x) = \frac{1}{e^{\phi(x)}} \frac{d\Psi(x)}{dx} = \frac{1}{\Psi(x)} \frac{d\Psi(x)}{dx} = \frac{\Psi'(x)}{\Psi(x)} \qquad (9.6)$$

Substitution in Eq. (9.5) yields

$$\{\frac{d\psi(x)}{dx} + \psi^2(x) + A(x)\,\psi(x) + B(E,x)\}\,e^{\phi(x)} = 0$$

which leads to

$$\frac{d\psi(x)}{dx} = -B(E,x) - A(x)\,\psi(x) - \psi^2(x)$$

which is a Riccati equation [see Eq. (8.4)].

The logarithmic transformation fails, of course, if $\Psi(x)$ presents nodes (see Aharonov and Au 1979) and an alternative may be found in an angular analogue of the Riccati equation (Prüfer 1926, Milne 1930, Drukarev 1949, Francetti 1957). The procedure will be illustrated here on the basis of the work of Mielnik and Reyes (1996) for the classical analogue of the 1-dimensional Schrödinger equation.

Suppose that we substitute t for x and denote it as time. Using the notation

$$q = \Psi(t)$$

$$p = \frac{d\Psi(t)}{dt} = \frac{dq}{dt} \qquad\qquad (9.7a)$$

Eq. (9.3) becomes (with $\hbar = m = 1$)

$$\frac{dp}{dt} = 2[V(t) - E]q \qquad\qquad (9.7b)$$

Equations (9.7) are the canonical equations corresponding to the Hamiltonian function

$$H(t) = \frac{1}{2}p^2 + [E - V(t)]q$$

Introducing the transformation

$$q = \rho\cos\alpha \qquad\qquad p = \rho\sin\alpha$$

the above canonical equations become (using the standard notation for differentiation with respect to time)

$$\dot{\rho}\cos\alpha - \dot{\alpha}\,\rho\sin\alpha = \rho\sin\alpha$$

$$\dot{\rho}\sin\alpha + \dot{\alpha}\,\rho\cos\alpha = 2[V(t) - E]\,\rho\cos\alpha$$

Respective multiplication of these equations with $-\sin \alpha$ and $\cos \alpha$ and adding the resulting equations yields (after elimination of ρ)

$$\dot{\alpha} = 2[V(t) - E] \cos^2 \alpha - \sin^2 \alpha \qquad (9.8)$$

which may be written as

$$\frac{\dot{\alpha}}{\cos^2 \alpha} = 2[V(t) - E] - \frac{\sin^2 \alpha}{\cos^2 \alpha}$$

and finally as

$$\frac{d \tan \alpha}{dt} = 2[V(t) - E] - \tan^2 \alpha$$

which is the angular analogue of the Riccati equation. The advantage of Eq. (9.8) is that it can be solved for arbitrary values of α without any problem of singularities.

9.2 Solution of Sturm-Liouville Equations

We will now examine the series expansion of two types of Sturm-Liouville equations *via* the Riccati equation for the logarithmic derivative.

The Riccati-Pade method, in particular, has been applied to the study of both separable and non-separable quantum-mechanical problems. It is based on the representation of the logarithmic derivative of the eigenfunction by means of a Pade approximant and the eigenvalues are obtained as roots of a Hankel determinant constructed with the Taylor-coefficients of the function (Fernandez *et al.* 1989, Fernandez 1992, Fernandez and Guardiola 1993, Fernandez *et al.* 1993, Fernandez 1995ab, Fernandez 1996, and Fernandez and Guardiola 1997).

9.2.1 Equations $\Psi''(E,x) + A(x)\Psi'(E,x) + B(E,x)\Psi(E,x) = 0$

This equation, which is of interest in the study of the Hydrogen molecular ion, has been treated by Fernandez (1995*a*), using the Riccati-Pade method, with the following conditions: $\Psi(E,x)$ is quadratically integrable and $A(x)$ and $B(E,x)$ are regular (but may exhibit poles at the endpoints) in the interval (a,b), with the assumption that $a \leq 0 < b$.

Instead of the simple logarithmic derivative [see Eq. (9.6)], the regularized expression

$$\psi(x) = \frac{s}{x} - \frac{\Psi'(E,x)}{\Psi(E,x)}$$

was used; s is a constant, whose value is chosen so that any poles of $\Psi'(E,x)/\Psi(E,x)$ at $x = 0$ are eliminated (see below). Simple manipulation yields in this case

$$\psi'(E,x) = \{\frac{s(s-1)}{x^2} + A(x)\frac{s}{x} + B(E,x)\} - \{\frac{2s}{x} + A(x)\}\psi(E,x) + \psi^2(E,x) \qquad (9.9)$$

Taking into account that $\psi(x)$ is regular at $x = 0$, it may be expanded

$$\psi(E,x) = \sum_{j=-1}^{\infty} f_j(E)\, x^{j+1} \qquad (9.10)$$

by a Taylor series, whose coefficients $f_j(E)$ may be obtained from those of the Laurent expansions (see, e.g., Mathews and Walker 1970)

$$A(x) = \sum_{j=-1}^{\infty} a_j\, x^j \qquad\qquad B(E,x) = \sum_{j=-2}^{\infty} b_j(E)\, x^j \qquad (9.11)$$

as follows. Substitution of Eqs. (9.10) and (9.11) into Eq. (9.9) yields

$$\sum_{j=0}^{\infty} (j+1)f_j(E)x^j = \{\frac{s(s-1)}{x^2} + \frac{s}{x}\sum_{j=-1}^{\infty} a_j x^j + \sum_{j=-2}^{\infty} b_j(E)x^j\}$$
$$- \{\frac{2s}{x} + \sum_{j=-1}^{\infty} a_j x^j\}\sum_{k=-1}^{\infty} f_k(E)x^{k+1} + \sum_{j=-1}^{\infty}\sum_{k=-1}^{\infty} f_j(E)f_k(E)x^{j+k+2} \qquad (9.12)$$

The expansion on the *lhs* and the expansion on the *rhs* of this equation must yield the same value for any arbitrary x within the interval considered. This condition will be satisfied if the coefficients of every power of x are identical on both sides of the equation. Thus one obtains

for x^{-2} $\qquad 0 = s(s - 1) + s\, a_{-1} + b_{-2}(E)$ $\qquad\qquad\qquad\qquad$ (9.13a)

for x^{-1} $\qquad 0 = s\, a_0 + b_{-1}(E) - (2s + a_{-1})f_{-1}(E)$ $\qquad\qquad$ (9.13b)

for x^0 $\qquad f_0(E) = sa_1 + b_0(E) - (2s + a_{-1})f_0(E) - a_0 f_{-1}(E) + f_{-1}^2(E)$ $\qquad$ (9.13c)

for x^1 $\qquad 2f_1(E) = sa_2 + b_1(E) - (2s + a_{-1})f_1(E) - a_0 f_0(E) - a_1 f_{-1}(E)$

$$\qquad\qquad\qquad + 2f_{-1}(E)f_0(E) \qquad\qquad\qquad\qquad\qquad (9.13d)$$

. . .

for x^n $(n + 1)f_n(E) = sa_{n+1} + b_n(E) - 2sf_n(E) - \sum\limits_{j=-1}^{\infty} a_j f_{n-j-1}(E)$ (9.13e)

$$+ \sum\limits_{j=-1}^{\infty} f_j(E)f_{n-j-2}(E)$$

Equations (9.13a) and (9.13b) are then used to obtain the values of s and $f_{-1}(E)$ that will remove the poles at the origin. That is: solution of the quadratic equation

$$s^2 + (a_{-1} - 1)s + b_{-2}(E) = 0$$

yields the values of s and then $f_{-1}(E)$ is obtained from

$$f_{-1}(E) = \frac{sa_0 + b_{-1}(E)}{2s + a_{-1}}$$

(with the understanding that the choice of the value of s to be used is to be made on the basis of the physical conditions for the actual problem under consideration). From the remaining Eqs. (9.13) one obtains then

$$f_0(E) = \frac{sa_1 + b_0(E) - a_0 f_{-1}(E) + f^2_{-1}(E)}{2s + a_{-1} + 1}$$

$$f_1(E) = \frac{sa_2 + b_1(E) - a_0 f_0(E) - a_1 f_{-1}(E) + 2f_{-1}(E)f_0(E)}{2s + a_{-1} + 2}$$

. . .

$$f_n(E) = \frac{sa_{n+1} + b_n(E) - \sum\limits_{j=0}^{\infty} a_j f_{n-j-1}(E) + \sum\limits_{j=-1}^{\infty} f_j(E)f_{n-j-2}(E)}{2s + a_{-1} + (n+1)}$$

This last expression, which gives the general recursion relationship for the determination of $f_n(E)$, may be rewritten as

$$f_n(E) = \frac{sa_{n+1} + b_n(E) + \sum\limits_{j=0}^{\infty} [f_{j-1}(E) - a_j]f_{n-j-1}(E)}{2s + a_{-1} + (n+1)}$$

The above results are eigenvalue-dependent. That is: the function $\psi(E,x)$, and therefore the corresponding coefficients $f_j(E)$, depend on the eigenvalue E, through the coefficients $b_j(E)$ of the function $B(E,x)$. Estimates of the eigenvalues may be obtained by a procedure supported by the results of numerical calculations, even though no rigorous proof has been given for it.

A Padé approximant (see, e.g., Press *et al.* 1992) could be found for the Taylor expansion, Eq. (9.10). This approximant would yield exactly the coefficients $f_j(E)$ for $j = 0, 1, ..., 2n + d$, if the degrees of its numerator and denominator were $n + d$ and n, respectively. In fact, if the approximant were an exact solution of Eq. (9.9) for a given eigenvalue, it would yield all the coefficients $f_j(E)$ of the function $\psi(E,x)$ corresponding to that eigenvalue. It has been observed numerically that estimates of the eigenvalues may be obtained if the Padé approximant also yields the coefficient $f_{2n+d+1}(E)$, those eigenvalues being the roots of the Hankel determinant.

$$H_{n+1}^d(E) = \begin{vmatrix} f_{d+1} & f_{d+2} & \cdots & f_{n+d+1} \\ f_{d+2} & f_{d+3} & \cdots & f_{n+d+2} \\ \cdots & \cdots & \cdots & \cdots \\ f_{n+d+1} & f_{n+d+2} & \cdots & f_{2n+d+1} \end{vmatrix} \qquad (9.14)$$

where the E-dependence of the coefficients $f_j(E)$ has been omitted for simplicity (Fernandez *et al.* 1989, Fernandez 1992, Fernandez *et al.* 1993, Fernandez and Guardiola 1993). This formulation has been applied (Fernandez 1995) to the case of the Hydrogen-ion molecule.

An alternate procedure (Fernandez *et al.* 1989a), applied to a 1-dimensional Schrödinger equation with $V(x) = - Z/x + \lambda x^2$ (where Z is the nuclear charge and λ is an arbitrary parameter), is equivalent in certain instances to a non-perturbative method (Hall 1977, Silva and Canuto 1982, 1984). This formulation has been applied to the study of the Zeeman effect in Hydrogen. [See the work of Fernandez *et al.* (1989a) and of Aharonov and Au (1979), Eletsky and Popov (1980), Avron (1981), and Privman (1981) for details of the series expansion for the eigenvalues.]

9.2.2 Equations $\Psi''(E,x) + B(E,x)\Psi(E,x) = 0$

In this case the equations to be solved [see Section 9.1.3] are

$$\Psi''(E,x) + B(E,x)\Psi(E,x) = 0 \qquad (9.15a)$$

$$\psi'(E,x) = - B(E,x) - \psi^2(E,x) \qquad (9.15b)$$

with

$$B(E,x) = E - V(x)$$

Asymptotic (Erdelyi 1956, Atkinson 1957, Hochstadt 1961, Fix 1967, Fulton 1982, Atkinson and Fulton 1984) and exact (Harris 1989) solutions have been obtained for Eq. (9.15b) for (large) positive values of E but here we will centre our attention on those cases involving potentials of possible quantum-chemical interest.

Solvable Hamiltonians are those for which the possibility exists of knowing their complete set of eigenfunctions as well as their eigenvalue spectrum (Infeld and Hull 1951; see also Abraham and Moses 1980, Luban and Pursey 1986, Pursey 1986,1987) while partially solvable and quasiexactly solvable potentials are those for which either one state or a family of states is exactly known (Killinbeck 1978, Flessas 1979, Flessas and Das 1980, Flessas 1981, 1982, Magyari 1981, Varma 1981, Saxena and Varma 1982, Whitehead *et al.* 1982, Bessis *et al.* 1987, Roy and Roychoudhury 1987, Dutta and Wiley 1988, Gallas 1988, Roychoudhury and Varshni 1988, de Sousa Doutra 1988, Leach *et al.* 1989).

In some instances, modelling of a physical problem leads to equations, Eq. (9.15a), which are, or are amenable to be exactly solvable with an additional perturbation. Perturbation theory, the logarithmic perturbation method (Imbo and Sukhatme 1984), and the perturbed ladder operator method (Bessis and Bessis 1990, 1991, 1992, 1994) have been used for the solution of the problem, but recently methods making use of a Riccati equation, Eq. (9.15b), have been proposed (Salem and Montemayor 1991, Bessis and Bessis 1995, 1996, Berrondo and Recamier 1997).

Bessis and Bessis (1995, 1996) solved the problem through the use of power series of functions appropriate for the case under study (symmetric anharmonic oscillator, perturbed Morse oscillator, singular anharmonic oscillator, and curved- and flat-space isotropic oscillator).

The work of Salem and Montemayor (1991) is concerned with the investigation, using a modified Riccati equation, of the partial solvability of Laurent-series-type (anharmonic polynomial oscillators, Coulomb plus polynomial terms, generalized Lennard-Jones-type) potentials as well as, in a generalized form, of the partial and quasiexact solvability of a larger class of potentials. The assumption is made that the potential is regular almost everywhere in the complex plane, meaning that its singularities, if any, will be found at the boundaries or at $|x| \rightarrow \infty$.

The function $\Psi_m(x)$ for the m-th excited state of the 1-dimensional Schrödinger equation has m (almost everywhere) simple zeros in the interval (a,b) on the real axis (Ince 1956, Dunham 1929, Leacock and Padgett 1983) and they can be isolated; that is, one can write

$$\Psi(x;n) = \varphi(x) \prod_{j=1}^{n} (x - x_j) \qquad n \geq m \qquad (9.16)$$

where n is the total number of zeros ($n \geq m$) and $\varphi(x)$ gives the leading behaviour of the function at the boundaries and is regular almost everywhere. Therefore for the logarithmic derivative one can write

$$\psi(x;n) = \phi(x) + \sum_{j=1}^{n} \frac{1}{x - x_j}$$

where

$$\phi(x) = \frac{d \ln \varphi(x)}{dx}$$

is also regular almost everywhere in the complex plane. Substitution of $\psi(x;n)$ into the Riccati equation, Eq. (9.15b), yields

$$\phi'(x) - \sum_{j=1}^{n} \frac{1}{(x - x_j)^2} = V(x) - E_m - \{\phi^2(x) + 2\phi(x) \sum_{j=1}^{n} \frac{1}{x - x_j}$$
$$+ \sum_{j=1}^{n} \sum_{k=1}^{n} \frac{1}{(x - x_j)(x - x_k)}\}$$

which is transformed into

$$\phi'(x) = V(x) - E_m + \sum_{j=1}^{n} \frac{1}{x - x_j} \{\frac{1}{x - x_j} - \sum_{k=1}^{n} \frac{1}{x - x_k}\} - 2\phi(x) \sum_{j=1}^{n} \frac{1}{x - x_j} - \phi^2(x)$$
$$= V(x) - E_m - \sum_{j \neq k=1}^{n} \frac{1}{(x - x_j)(x - x_k)} - 2\phi(x) \sum_{j=1}^{n} \frac{1}{x - x_j} - \phi^2(x) \qquad (9.17)$$

Taking into account that

$$\sum_{j \neq k=1}^{n} \frac{1}{(x - x_j)(x - x_k)} = \sum_{j \neq k=1}^{n} \{\frac{1}{x - x_k} - \frac{1}{x - x_j}\} \frac{1}{x_k - x_j}$$
$$= - \sum_{j \neq k=1}^{n} \frac{1}{(x - x_j)(x_k - x_j)} + \sum_{j \neq k=1}^{n} \frac{1}{(x - x_j)(x_j - x_k)}$$
$$= -2 \sum_{j \neq k=1}^{n} \frac{1}{(x - x_j)(x_k - x_j)}$$
$$= -2 \sum_{j=1}^{n} \frac{\gamma_j(n)}{x - x_j}$$

with

$$\gamma_j(n) = \sum_{k \neq j=1}^{n} \frac{1}{x_k - x_j} \qquad (9.18)$$

Eq. (9.17) becomes

$$\phi'(x) = V(x) - E_m - 2 \sum_{j=1}^{n} \frac{\phi(x) - \gamma_j(n)}{x - x_j} - \phi^2(x)$$

$$= V(x) - E_m - 2 F_m(x;n) - \phi^2(x) \qquad (9.19)$$

with

$$F_m(x;n) = \sum_{j=1}^{n} \frac{\phi(x) - \gamma_j(n)}{x - x_j}$$

Rewriting Eq. (9.19) as

$$\phi'(x) + 2F_m(x;n) + \phi^2(x) = V(x) - E_m \qquad (9.20)$$

one can offer the following argument. The potential $V(x)$, which is regular almost everywhere, as well as $\phi(x)$ and $\phi'(x)$, will not present singularities at $x = x_j$. It follows then that $F_m(x;n)$ will not have singularities at $x = x_j$ either. This condition will be satisfied if $\{\phi(x) - \gamma_j(n)\}$ is proportional to $(x - x_j)$. One can write

$$\phi(x) - \gamma_j(n) = c(x - x_j)$$

which, for $x = x_j$, yields

$$\gamma_j(n) = \phi(x_j) \qquad (9.21)$$

so that finally

$$F_m(x;n) = \sum_{j=1}^{n} \frac{\phi(x) - \phi(x_j)}{x - x_j} \qquad (9.22)$$

Taking into account Eq. (9.18) one can see that Eq. (9.21) really stands for a system of non-linear equations,

$$\sum_{j \neq k=1}^{n} \frac{1}{x_k - x_j} = \phi(x_j) \qquad (9.23)$$

This system of equations can have more than one set of solutions $\{x_n\}$, with the result that more than one function with the same number of zeros may be obtained. For a given n, if the $F_m(x;n)$ for different sets of $\{x_n\}$ differ in their x-dependence, the associated functions correspond to different partially solvable potentials but if they differ by an additive constant the functions have different number of nodes and correspond to different states and the potential is quasiexactly solvable.

Thus, once the function $\phi(x)$ is known, being required for the solution of the system of non-linear equations, Eqs. (9.23), a family of functions with a specified number of zeros may be obtained. For example, for a finite Laurent-series-type potential

$$V(x) = \sum_{K=-2L}^{2M} v_K x^K \qquad\qquad L, M \geq 0$$

Salem and Montemayor (1991) find that the appropriate function $\phi(x)$ is given also by a finite Laurent series

$$\phi(x) = \sum_{K=-L}^{M} f_K x^K$$

so that, taking into account that $\phi(x)$ is the logarithmic derivative of $\varphi(x)$, one can write

$$\ell n\ \varphi(x) = \int \phi(x)dx = \int \{\sum_{K=-L}^{M} f_K x^K\}dx = \int \{\sum_{K=-L}^{M}{}' f_K x^K + \frac{f_{-1}}{x}\}dx$$

$$= \sum_{K=-L}^{M}{}' \frac{f_K}{K+1} x^{K+1} + \ell n\ x^{f_{-1}}$$

(where the primed summation indicates that $K \neq -1$) and finally

$$\varphi(x) \propto x^{f_{-1}} \exp\{\sum_{K=-L}^{M}{}' \frac{f_K}{K+1} x^{K+1}\}$$

The reader is referred to the work of Salem and Montemayor (1991) for additional details and, particularly, for the application of this formulation to other types of potentials.

Equation (9.15a) has also been treated by Fernandez and Guardiola (1997), using the Riccati-Pade method (see Section 9.2.1), with extension of the formulation to coupled-channel equations, appearing in the study of non-separable problems, of the form

$$\begin{pmatrix} \mathcal{H}_{11} & \mathcal{H}_{12} \\ \mathcal{H}_{21} & \mathcal{H}_{22} \end{pmatrix} \begin{pmatrix} \Psi_1(x) \\ \Psi_2(x) \end{pmatrix} = \begin{pmatrix} 0 \\ 0 \end{pmatrix}$$

with

$$\mathcal{H}_{ii} = -\frac{\partial^2}{\partial x^2} + B_{ii}(E, x)$$

$$\mathcal{H}_{ij} = \mathcal{H}_{ji} = B_{ij}(x)$$

The formulation has been applied to the study of coupled harmonic oscillators, a Coulomb-like problem, the Zeeman effect in Hydrogen, and a system with no bound states.

The Schrödinger-Riccati Equation

The Riccati equation also appears in the study of the Schrödinger equation when considering the pseudo local energies associated with the components (starting trial function plus the appropriate correction) of a Schrödinger eigenfunction: the variation of those pseudo local energies under a variation of the starting function is governed by corresponding Riccati equations.

The theoretical development leads to a series expansion, in powers of the correction function and with coefficients to be obtained from the starting function. For practical applications, the series expansion may be truncated to the appropriate finite degree, yielding an algebraic equation for the correction function. This algebraic equation, denoted as the Schrödinger-Riccati equation, may then be solved at each point of the electron configuration space.

The derivation of the Schrödinger-Riccati equation is presented in Chapter 10 while the practical details of the calculations and the characteristics of the results are illustrated in Chapter 11 for some chosen examples.

10 The Schrödinger-Riccati Equation

The derivation of the Schrödinger-Riccati equation for a spin-free Hamiltonian operator is best developed if the case of spin-free functions is discussed first, followed by the formulation for the evaluation of the energy and some practical considerations. Then the formulation may be easily generalized to the case of antisymmetric functions for many-electron systems (Fraga and Fraga 1998*ab*).

10.1 Formulation for Spin-Free Functions

The spin-free eigenfunction Ψ (not necessarily normalized), corresponding to an eigenvalue E of the Schrödinger equation for a spin-free Hamiltonian operator will be expressed as

$$\Psi = \phi + \varphi \tag{10.1a}$$

where ϕ is a known function (not necessarily normalized) and φ represents the correction function needed in order to obtain the eigenfunction Ψ. Within the framework of this approach, the goal is then the determination of the function φ, to be achieved by means of the Schrödinger-Riccati equation.

The function ϕ may be normalized or not and the respective notation ϕ_N and ϕ_U will be used, with the relationship $\phi_N = k\phi_U$, where k is the normalization constant. Correspondingly, Eq. (10.1a) should be written as

$$\Psi_{(N)} = \phi_N + \varphi_{(N)} \qquad \text{or} \qquad \Psi_{(U)} = \phi_U + \varphi_{(U)} \tag{10.1b}$$

where the subscripts (N) and (U), used for Ψ and φ, serve to indicate that they have been obtained using ϕ_N or ϕ_U, respectively. The specification of whether ϕ is normalized or not will be omitted throughout the derivation for simplicity, except where needed. [See, in particular, Sections 10.1.4 and 10.1.5.]

10.1.1 Basic Definitions and Relationships

The function ϕ will be assumed to contain a variational parameter, whose infinitesimal variation (under the condition that $\Delta\Psi = 0$ and, consequently, $\Delta E = 0$), will result in the infinitesimal variations $\Delta\phi$ and $\Delta\varphi = -\Delta\phi$. The formulation will make use of the differentiation of some basic quantities and of the Schrödinger equation with respect to ϕ.

The basic quantities to be considered are $\mathcal{H}\phi$, hereafter denoted as $H^{(0)}$, and

$$S^{(0)} = H^{(0)} - E\phi = (\mathcal{H} - E)\phi = -(\mathcal{H} - E)\varphi \tag{10.2}$$

which is obtained, after reordering, by substitution of Eq. (10.1) into the Schrödinger equation. For the differentiation of these quantities with respect to ϕ we will use the notation

$$H^{(n)} = \frac{\partial^n \mathcal{H}\phi}{\partial\phi^n} \qquad S^{(1)} = H^{(1)} - E \qquad S^{(n)} = H^{(n)} \quad \text{(for } n > 1\text{)} \tag{10.3}$$

The quantities $H^{(n)}$ and $S^{(n)}$, evaluated for ϕ_N and ϕ_U, are related by

$$H^{(n)}_{(N)} = k^{-(n-1)} H^{(n)}_{(U)} \qquad S^{(n)}_{(N)} = k^{-(n-1)} S^{(n)}_{(U)} \tag{10.4}$$

Differentiation of the Schrödinger equation with respect to ϕ yields

$$\frac{\partial \mathcal{H}\phi}{\partial\phi} + \frac{\partial \mathcal{H}\varphi}{\partial\phi} = 0$$

which may be rewritten as

$$H^{(1)} = \frac{\partial \mathcal{H}\phi}{\partial\phi} = -\frac{\partial \mathcal{H}\varphi}{\partial\phi} = -\frac{\partial \mathcal{H}\varphi}{\partial\varphi}\frac{\partial\varphi}{\partial\phi} = \frac{\partial \mathcal{H}\varphi}{\partial\varphi} \tag{10.5}$$

Using the local energies (see Chapter 5)

$$\varepsilon_\phi = \frac{\mathcal{H}\phi}{\phi} \qquad \varepsilon_\varphi = \frac{\mathcal{H}\varphi}{\varphi} \tag{10.6}$$

Eq. (10.5) becomes

$$H^{(1)} = \varepsilon_\phi + \phi\frac{\partial\varepsilon_\phi}{\partial\phi} = \varepsilon_\phi + \phi\varepsilon_\phi^{(1)} \tag{10.7a}$$

$$H^{(1)} = \varepsilon_\varphi + \varphi\frac{\partial\varepsilon_\varphi}{\partial\varphi} = \varepsilon_\varphi - \varphi\frac{\partial\varepsilon_\varphi}{\partial\phi} = \varepsilon_\varphi - \varphi\varepsilon_\varphi^{(1)} \tag{10.7b}$$

where $\varepsilon_\phi^{(1)}$ and $\varepsilon_\varphi^{(1)}$ denote, respectively, the derivatives of ε_ϕ and ε_φ with respect to ϕ.

10.1.2 The Riccati Local Energies

Operating on Ψ, Eq. (10.1), with $(\mathcal{H} - \varepsilon)$, where ε is a variable with energy dimensions, yields

$$(\mathcal{H} - \varepsilon)\Psi = (E - \varepsilon)\Psi = (\mathcal{H} - \varepsilon)\phi + (\mathcal{H} - \varepsilon)\varphi$$

which will be rewritten as

$$\Psi = F(\varepsilon) + f(\varepsilon)$$

with

$$F(\varepsilon) = \frac{\mathcal{H}\phi - \varepsilon\phi}{E - \varepsilon} = \frac{\varepsilon_\phi - \varepsilon}{E - \varepsilon}\phi \qquad f(\varepsilon) = \frac{\mathcal{H}\varphi - \varepsilon\varphi}{E - \varepsilon} = \frac{\varepsilon_\varphi - \varepsilon}{E - \varepsilon}\varphi$$

from which one can define the additional functions

$$G(\varepsilon) = F(\varepsilon) - \phi = \frac{\mathcal{H}\phi - E\phi}{E - \varepsilon} = \frac{S^{(0)}}{E - \varepsilon}$$

$$g(\varepsilon) = f(\varepsilon) - \varphi = \frac{\mathcal{H}\varphi - E\varphi}{E - \varepsilon} = \frac{-S^{(0)}}{E - \varepsilon}$$

The values of the above functions for $\varepsilon = \varepsilon_\phi$ and $\varepsilon = \varepsilon_\varphi$ are

$$F(\varepsilon_\phi) = 0 \qquad\qquad F(\varepsilon_\varphi) = \frac{\varepsilon_\phi - \varepsilon_\varphi}{E - \varepsilon_\varphi}\phi \equiv \Psi \qquad\qquad (10.8a)$$

$$f(\varepsilon_\phi) = \frac{\varepsilon_\varphi - \varepsilon_\phi}{E - \varepsilon_\phi}\varphi \equiv \Psi \qquad f(\varepsilon_\varphi) = 0 \qquad\qquad (10.8b)$$

$$G(\varepsilon_\phi) = -\phi \qquad\qquad G(\varepsilon_\varphi) = \frac{S^{(0)}}{E - \varepsilon_\varphi} \equiv \varphi \qquad\qquad (10.8c)$$

$$g(\varepsilon_\phi) = \frac{-S^{(0)}}{E - \varepsilon_\phi} \equiv \phi \qquad g(\varepsilon_\varphi) = -\varphi \qquad\qquad (10.8d)$$

Figure 10.1 presents, in part, the behaviour of some of these functions for a point of the electron configuration space at which $S^{(0)} < 0$.

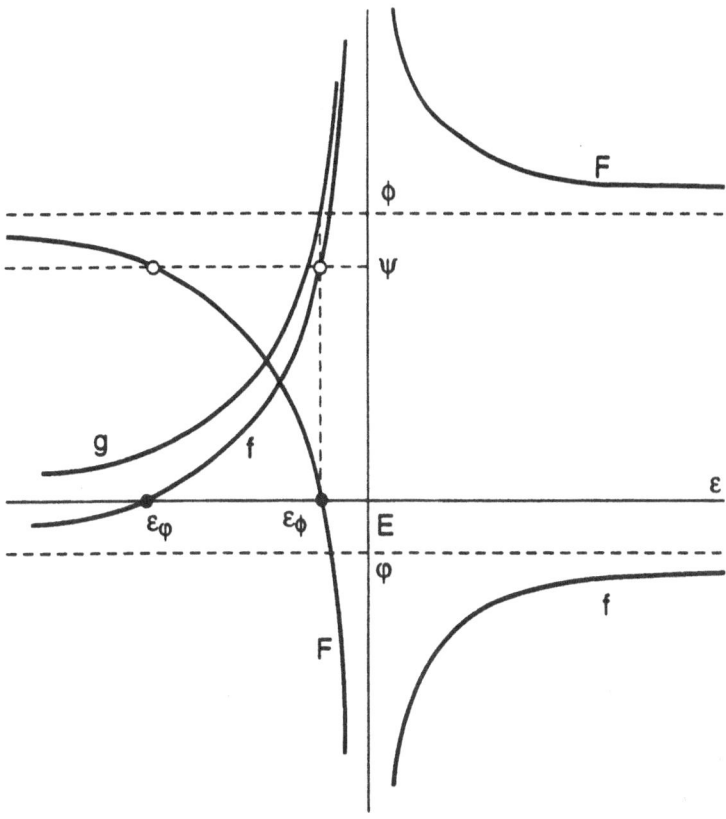

Figure 10.1. Schematic representation of the functions $F(\varepsilon)$, $f(\varepsilon)$, and $g(\varepsilon)$ at a *PECS* for which $S^{(0)} < 0$. For simplicity, only the branch of $g(\varepsilon)$, which is of interest for the discussion, has been represented.

Equations (10.7) may now be rewritten, using the definitions of ϕ and φ given by Eqs (10.8d) and (10.8c), respectively, as

$$H^{(1)} = \varepsilon_\phi - \frac{S^{(0)}}{E - \varepsilon_\phi}\, \varepsilon_\phi^{(1)} \qquad\qquad H^{(1)} = \varepsilon_\varphi - \frac{S^{(0)}}{E - \varepsilon_\varphi}\, \varepsilon_\varphi^{(1)}$$

or

$$\varepsilon_\phi^{(1)} = \frac{1}{S^{(0)}}\, (E - \varepsilon_\phi)(\varepsilon_\phi - H^{(1)}) = -\frac{EH^{(1)}}{S^{(0)}} + \frac{E + H^{(1)}}{S^{(0)}}\, \varepsilon_\phi - \frac{1}{S^{(0)}}\, \varepsilon_\phi^2 \qquad (10.9a)$$

$$\varepsilon_\varphi^{(1)} = \frac{1}{S^{(0)}}\, (E - \varepsilon_\varphi)(\varepsilon_\varphi - H^{(1)}) = -\frac{EH^{(1)}}{S^{(0)}} + \frac{E + H^{(1)}}{S^{(0)}}\, \varepsilon_\varphi - \frac{1}{S^{(0)}}\, \varepsilon_\varphi^2 \qquad (10.9b)$$

which are the Riccati equations for ε_ϕ and ε_φ, respectively, with [see Eq. (8.4)]

$$P(\phi) = -\frac{EH^{(1)}}{S^{(0)}} \qquad Q(\phi) = \frac{E + H^{(1)}}{S^{(0)}} \qquad R(\phi) = -\frac{1}{S^{(0)}}$$

Therefore ε_ϕ and ε_φ will be denoted afterwards as the Riccati local energies. Successive differentiation of Eq. (10.9) with respect to ϕ yields

$$\varepsilon^{(n)} = \frac{1}{S^{(0)}} (E - \varepsilon)(n\varepsilon^{(n-1)} - H^{(n)}) \tag{10.10}$$

where ε stands for either ε_ϕ or ε_φ.

10.1.3 The Schrödinger-Riccati Equation

Let us now choose a function ϕ and generate a new function, $\phi' = \phi + \Delta\phi$, by an infinitesimal variation of ϕ. For simplicity in the notation we will represent the corresponding Riccati local energies, ε_φ and ε'_φ, by ε and ε', respectively. The value of the function Ψ, at the *PECS* under consideration, will be given indistinctly [see Eq. 10.8a)] by

$$\Psi = \frac{\mathcal{H}\phi - \phi\varepsilon}{E - \varepsilon} \equiv \frac{\mathcal{H}\phi' - \phi'\varepsilon'}{E - \varepsilon'}$$

from which one obtains

$$\varepsilon' = \frac{\varepsilon S^{(0)} - (E - \varepsilon)\mathcal{H}(\Delta\phi)}{S^{(0)} - (E - \varepsilon)(\Delta\phi)} \tag{10.11}$$

with

$$\mathcal{H}(\Delta\phi) = \mathcal{H}\phi' - \mathcal{H}\phi$$

One may also express ε', as a function of ϕ, in terms of ε and of its derivatives, obtained using Eq. (10.10), by the Taylor expansion

$$\varepsilon' = \varepsilon + \varepsilon^{(1)} \frac{(\Delta\phi)}{1!} + \varepsilon^{(2)} \frac{(\Delta\phi)^2}{2!} + \varepsilon^{(3)} \frac{(\Delta\phi)^3}{3!} + \dots$$

which can be equated to Eq. (10.11) to yield, after simplification,

$$\mathcal{H}(\Delta\phi) - \varepsilon(\Delta\phi) + \frac{1}{S^{(0)}} [S^{(0)} - (E - \varepsilon)(\Delta\phi)]\{(\varepsilon - H^{(1)}) \frac{(\Delta\phi)}{1!} + (2\varepsilon^{(1)} - H^{(2)}) \frac{(\Delta\phi)^2}{2!} + \dots\} = 0$$

This equation may be used to express ε to successive orders. When including the first, second, and third terms of the expansion in the curly brackets one obtains, respectively,

$$\frac{(\Delta\phi)^2}{S^{(0)}}\{\varepsilon^2 - (E + H^{(1)})\varepsilon + EH^{(1)}\} + \{\mathcal{H}(\Delta\phi) - H^{(1)}\frac{(\Delta\phi)}{1!}\} = 0$$

$$\frac{(\Delta\phi)^3}{S^{(0)}}\{-\frac{1}{S^{(0)}}\varepsilon^3 + \frac{1}{S^{(0)}}(2E + H^{(1)})\varepsilon^2 - [\frac{1}{S^{(0)}}E(E + 2H^{(1)}) + \frac{1}{2!}H^{(2)}]\varepsilon$$

$$+ E(\frac{1}{S^{(0)}}EH^{(1)} + \frac{1}{2!}H^{(2)})\} + \{\mathcal{H}(\Delta\phi) - H^{(1)}\frac{(\Delta\phi)}{1!} - H^{(2)}\frac{(\Delta\phi)^2}{2!}\} = 0$$

$$\frac{(\Delta\phi)^4}{S^{(0)}}\{\frac{1}{\left(S^{(0)}\right)^2}\varepsilon^4 - \frac{1}{\left(S^{(0)}\right)^2}(3E + H^{(1)})\varepsilon^3 + [\frac{3}{\left(S^{(0)}\right)^2}E(E + H^{(1)}) + \frac{1}{2!S^{(0)}}H^{(2)}]\varepsilon^2$$

$$- [\frac{1}{\left(S^{(0)}\right)^2}E^2(E + 3H^{(1)}) + \frac{1}{S^{(0)}}EH^{(2)} + \frac{1}{3!}H^{(3)}]\varepsilon$$

$$+ \frac{1}{\left(S^{(0)}\right)^2}E^3H^{(1)} + \frac{1}{2!S^{(0)}}H^{(2)} + \frac{1}{3!}EH^{(3)}\}$$

$$+ \{\mathcal{H}(\Delta\phi) - H^{(1)}\frac{(\Delta\phi)}{1!} - H^{(2)}\frac{(\Delta\phi)^2}{2!} - H^{(3)}\frac{(\Delta\phi)^3}{3!}\} = 0$$

The expressions within the second set of curly brackets in each of the above equations may be recognized as representing the error, dependent on the number of terms considered initially, in the expansion of

$$H\phi' = H^{(0)} + H^{(1)}\frac{(\Delta\phi)}{1!} + H^{(2)}\frac{(\Delta\phi)^2}{2!} + H^{(3)}\frac{(\Delta\phi)^3}{3!} + \dots$$

and therefore it may be assumed that, to a significantly high-order approximation, that error will vanish. The remaining terms in the above equations may be reexpressed as

$$\frac{1}{S^{(0)}}(E - \varepsilon)(\varepsilon - H^{(1)}) = 0$$

$$\frac{1}{S^{(0)}}(E - \varepsilon)(\varepsilon - H^{(1)}) - \frac{1}{2!}H^{(2)} = 0$$

$$\frac{1}{\left(S^{(0)}\right)^2}(E - \varepsilon)^2(\varepsilon - H^{(1)}) - \frac{1}{2!S^{(0)}}(E - \varepsilon)H^{(2)} - \frac{1}{3!}H^{(3)} = 0$$

or equivalently, as

$$\varepsilon^{(1)} = 0 \qquad \varepsilon^{(1)} = \frac{1}{2!} H^{(2)} \qquad \varepsilon^{(1)} = \frac{1}{2!} H^{(2)} + \frac{1}{3!} \frac{S^{(0)}}{E - \varepsilon} H^{(3)}$$

which might be obtained directly from the expressions for $\varepsilon^{(n)}$, Eq. (10.10), when assuming respectively that $\varepsilon^{(1)}$, $\varepsilon^{(2)}$, and $\varepsilon^{(3)}$ vanish. Generalization of this result yields

$$\varepsilon^{(1)} = \frac{1}{2!} H^{(2)} + \frac{1}{3!} \frac{S^{(0)}}{E - \varepsilon} H^{(3)} + \frac{1}{4!} \frac{(S^{(0)})^2}{(E - \varepsilon)^2} H^{(4)} + \dots$$

which becomes

$$\varepsilon^{(1)} = \frac{1}{2!} H^{(2)} + \frac{1}{3!} \varphi H^{(3)} + \frac{1}{4!} \varphi^2 H^{(4)} + \dots$$

when taking into account Eq. (10.8c) defining φ. Finally, taking into account Eq. (10.7b), one obtains

$$\varepsilon_\varphi = H^{(1)} + \frac{1}{2!} \varphi H^{(2)} + \frac{1}{3!} \varphi^2 H^{(3)} + \frac{1}{4!} \varphi^3 H^{(4)} + \dots$$

where the complete notation has been restored for ε_φ.

Substitution of this expression into Eq. (10.8a) yields

$$\Psi = \frac{(\varepsilon_\varphi - H^{(1)}) - \frac{1}{2!} \varphi H^{(2)} - \frac{1}{3!} \varphi^2 H^{(3)} - \dots}{(E - H^{(1)}) - \frac{1}{2!} \varphi H^{(2)} - \frac{1}{3!} \varphi^2 H^{(3)} - \dots} \phi$$

which may be rewritten as

$$\Psi = \frac{\phi}{E - H^{(1)}} \frac{\sum\limits_{n=0} a_n \varphi^n}{b_0 + \sum\limits_{n=1} b_n \varphi^n} \tag{10.12}$$

with

$$a_0 = \varepsilon_\varphi - H^{(1)} \qquad a_n = -\frac{1}{(n+1)!} H^{(n+1)} \qquad \text{(for } n > 0\text{)}$$

$$b_0 = 1 \qquad\qquad b_n = - \frac{1}{(n+1)!} \frac{H^{(n+1)}}{E - H^{(1)}} \qquad \text{(for } n > 0)$$

The fact that the second fraction in Eq. (10.12) is the Padé approximant to the series expansion $\Sigma\, c_n \varphi^n$ should be mentioned, in the expectation that its special convergence characteristics (Press *et al.* 1992) might be conserved when Eq. (10.12) is transformed.

The transformation of Eq. (10.12), taking into account Eq. (10.1), yields the Schrödinger-Riccati equation

$$\sum_{n=0}^{\infty} \frac{1}{n!} S^{(n)}\, \varphi^n = \sum_{n=0}^{\infty} \frac{1}{n!} S^{(n)}\, (\Psi - \phi)^n = 0 \qquad\qquad (10.13)$$

which shows that the original problem of solving the Schrödinger equation for Ψ and E has been transformed into the problem of finding Ψ and E (contained in $S^{(0)}$ and $S^{(1)}$) which satisfy this equation. This equation may be rewritten as

$$H^{(0)} - E\phi + H^{(1)}\varphi - E\varphi + \sum_{n=2}^{\infty} \frac{1}{n!} S^{(n)}\, (\Psi - \phi)^n = -E\Psi + \sum_{n=0}^{\infty} \frac{1}{n!} H^{(n)}\, (\Psi - \phi)^n = 0$$

or, equivalently,

$$E\Psi \equiv \mathcal{H}\Psi = \sum_{n=0}^{\infty} \frac{1}{n!} H^{(n)}\, (\Psi - \phi)^n \qquad\qquad (10.14)$$

which resembles a Taylor expansion for $\mathcal{H}\Psi$.

10.1.4 Evaluation of the Correction Function φ

In the practical application of the Schrödinger-Riccati equation for the evaluation of the correction function φ it will be necessary to truncate it

$$\sum_{n=0}^{t} \frac{1}{n!} S^{(n)}\, \varphi^n = 0 \qquad\qquad (10.15)$$

to its first t terms, with the corresponding truncation error, which will depend on t and on the quality of ϕ (for a given t). The resulting equation of degree $(t-1)$, if solvable (see below), will yield the value of φ at the *PECS* under consideration. In this approach it is assumed that the eigenvalue E is known but the procedure may also be used for its determination, as discussed below.

The actual solution of such an equation is customarily carried out after transforming it

$$\frac{t!}{S^{(t)}} \sum_{n=0}^{t} \frac{1}{n!} S^{(n)} \varphi^n = 0 \qquad (10.16)$$

so that the coefficient of the highest power of the variable is one. If it happens that $S^{(t)} = 0$ at a given *PECS*, it will be necessary to obtain the solution from the equation with degree $(t-1)$ or $(t+1)$.

For some *PECS* (in particular those with large values of the coordinates) the quantities $S^{(n)}$ may be very large and the expected values of φ may be rather small, with the result that the solution of the equation may be faced with a problem of precision. In such cases it will be appropriate to transform the equation, making use of the relationship between φ and ε_φ, Eq. (10.8c), into an equation

$$\sum_{p=0}^{n} (-1)^p \varepsilon_\varphi^{n-p} \sum_{q=0}^{p} \frac{(n-q)!}{(n-p)!(p-q)!q!} E^{p-q} (S^{(0)})^{q-1} S^{(q)} = 0$$

in powers of ε_φ.

For example, when using a quartic equation, the two possibilities are

$$\varphi^4 + 4 \frac{S^{(3)}}{S^{(4)}} \varphi^3 + 12 \frac{S^{(2)}}{S^{(4)}} \varphi^2 + 24 \frac{S^{(1)}}{S^{(4)}} \varphi + 24 \frac{S^{(0)}}{S^{(4)}} = 0 \qquad (10.17a)$$

$$\varepsilon_\varphi^4 - [4E - S^{(1)}]\varepsilon_\varphi^3 + [6E^2 + 3ES^{(1)} + \frac{1}{2!} S^{(0)}S^{(2)}]\varepsilon_\varphi^2$$

$$- [4E^3 + 3E^2S^{(1)} + ES^{(0)}S^{(2)} + \frac{1}{3!} (S^{(0)})^2 S^{(3)}]\varepsilon_\varphi$$

$$+ [E^4 + E^3S^{(1)} + \frac{1}{2!} E^2S^{(0)}S^{(2)} + \frac{1}{3!} (S^{(0)})^2 S^{(3)} + \frac{1}{4!} (S^{(0)})^3 S^{(4)}] = 0 \quad (10.17b)$$

The degree of the equation to be used should be decided on the basis of the quality of the starting function ϕ. A numerical approach to the solution of the equation is always possible but the analytical approach is only possible up to quartic equations and the corresponding problems to be encountered may be summarized as follows:

(a) The first-degree equation in φ will yield a very simple solution but it will be appropriate only if ϕ constitutes already a very good approximation to Ψ. This equation will fail at, and in the vicinity of, those *PECS* for which $S^{(0)} = 0$ or $S^{(1)} = 0$, respectively.

(b) The quadratic equation in φ, with solutions involving a square root, will fail for those *PECS* for which $(S^{(1)})^2 < 2S^{(0)}S^{(2)}$.

(c) The solution of the cubic and quartic equations involves the use and transformation of trigonometric functions.

It is regretable that the analytical approach, which would be highly desirable, will only be appropriate in a very limited number of cases. The problem with the numerical approach is that it will generate a wealth of data for the complete description of the electron configuration space in the case of many-electron systems.

At this point the dependence on the normalization of ϕ must be examined. Taking into account Eqs. (10.4) we can write

$$\sum_{n=0}^{t} \frac{1}{n!} S_{(N)}^{(n)} \varphi_{(N)}^n = \sum_{n=0}^{t} \frac{1}{n!} k^{-(n-1)} S_{(U)}^{(n)} \varphi_{(N)}^n = k \sum_{n=0}^{t} \frac{1}{n!} S_{(U)}^{(n)} \left(\frac{\varphi_{(N)}}{k}\right)^n$$

$$= k \sum_{n=0}^{t} \frac{1}{n!} S_{(U)}^{(n)} \varphi_{(U)}^n = 0$$

Solutions of the two equations

$$\sum_{n=0}^{t} \frac{1}{n!} S_{(N)}^{(n)} \varphi_{(N)}^n = 0 \qquad\qquad \sum_{n=0}^{t} \frac{1}{n!} S_{(U)}^{(n)} \varphi_{(N)}^n = 0$$

will yield the correction functions $\varphi_{(N)}$ and $\varphi_{(U)}$, respectively, related by

$$\varphi_{(N)} = k \, \varphi_{(U)}$$

Therefore, from Eqs. (10.1b) we obtain

$$\Psi_{(N)} = \phi_N + \varphi_{(N)} = k\phi_U + k\varphi_{(U)} = k(\phi_U + \varphi_{(U)}) = k\Psi_{(U)}$$

which shows that $\Psi_{(N)}$ and $\Psi_{(U)}$ differ by a constant factor k (which is the normalization constant of ϕ_U). On completion of the calculation, the resulting function, whether $\Psi_{(N)}$ or $\Psi_{(U)}$, may be normalized.

10.1.5 Evaluation of the Energy

As mentioned above, the Schrödinger-Riccati equation may also be used for the prediction of the energy.

As the energy is given by

$$E = \frac{1}{\Psi} (H^{(0)} + \mathcal{H}\varphi) \tag{10.18}$$

requiring a knowledge of both φ (contained in Ψ) and $\mathcal{H}\varphi$, a multistep procedure will be necessary, as follows.

At a given *PECS* the evaluation of φ, from an equation of the appropriate degree, is performed for a number M of input values of the energy, $E_{i(j)}$; $j = 1, 2, ..., M$, covering that interval of the energy spectrum where it is suspected that the true value lies. The calculation will yield M values φ_j, which are then used to evaluate the corresponding values Ψ_j and $\mathcal{H}\varphi_j$ (see below) and to obtain the corresponding M values of the predicted energies, $E_{p(j)}$. The search is continued, moving or narrowing (in order to increase the accuracy of the prediction) the energy window until the true eigenvalue is first bracketed and then finally found: the search is stopped when an input value, $E_{i(k)}$, is found for which $E_{p(k)} \equiv E_{i(k)}$, to the desired accuracy.

The evaluation of the numerical value of $\mathcal{H}\varphi$, needed in the above procedure, is performed as follows. Let us write (in Cartesian coordinates and atomic units)

$$\mathcal{H}\varphi = -\frac{1}{2} \sum_{i=1}^{N} \left\{ \frac{\partial^2 \varphi}{\partial x_i^2} + \frac{\partial^2 \varphi}{\partial y_i^2} + \frac{\partial^2 \varphi}{\partial z_i^2} \right\} + \mathcal{V}(x,y,z)\varphi \qquad (10.19)$$

where the summation runs over the N particles in the system and $\mathcal{V}(x,y,z)$ stands for a Coulombic potential. The term $\mathcal{V}(x,y,z)\varphi$ is obtained directly by multiplication of the numerical value of φ times the numerical value of $\mathcal{V}(x,y,z)$ at the *PECS* under consideration. The kinetic energy terms are obtained from the Schrödinger-Riccati equation by successive differentiation. One obtains, from Eq. (10.15),

$$\partial_i \varphi = -\frac{1}{T} \sum_{n=0}^{t} \frac{1}{n!} \varphi^n \partial_i S^{(n)} \qquad (10.20a)$$

$$\partial_i^2 \varphi = -\frac{1}{T} \left\{ \sum_{n=0}^{t} \frac{1}{n!} \varphi^n \partial_i^2 S^{(n)} + 2 \sum_{n=1}^{t} \frac{1}{(n-1)!} \varphi^{n-1} \partial_i S^{(n)} \partial_i \varphi \right. $$
$$\left. + \sum_{n=2}^{t} \frac{1}{(n-2)!} S^{(n)} \varphi^{n-2} (\partial_i \varphi)^2 \right\} \qquad (10.20b)$$

with

$$T = \sum_{n=1}^{t} \frac{1}{(n-1)!} S^{(n)} \varphi^{n-1} \qquad (10.20c)$$

and where $\partial_i \varphi$ and $\partial_i^2 \varphi$ stand for $\partial \varphi / \partial x_i$, $\partial \varphi / \partial y_i$, $\partial \varphi / \partial z_i$ and $\partial^2 \varphi / \partial x_i^2$, $\partial^2 \varphi / \partial y_i^2$, $\partial^2 \varphi / \partial z_i^2$, respectively.

It can be easily seen that the value obtained for E will not depend on whether one uses a normalized or an unnormalized function ϕ. Taking into account the relationships between $\varphi_{(N)}$ and $\varphi_{(U)}$ as well as between $\Psi_{(N)}$ and $\Psi_{(U)}$ we can write

$$E_{(N)} = \frac{1}{\Psi_{(N)}} (H_{(N)}^{(0)} + \mathcal{H}\varphi_{(N)}) = \frac{1}{k\Psi_{(U)}} (kH_{(U)}^{(0)} + k\mathcal{H}\varphi_{(U)}) = \frac{1}{\Psi_{(U)}} (H_{(U)}^{(0)} + \mathcal{H}\varphi_{(U)}) = E_{(U)}$$

which shows the identity of $E_{(N)}$ and $E_{(U)}$.

10.2 Formulation for Antisymmetric Functions

The preceding formulation may be extended in a straightforward manner to the case of antisymmetric functions for many-electron systems, containing a spin component. The computational cost in this case, however, will be greater, particularly for the determination of functions, as discussed below.

The details will be illustrated first for the two-electron case and then we will examine how to proceed in the general case and we will point out the difficulties to be faced.

10.2.1 Two-electron Systems

In the case when a single-determinant function constitutes an appropriate starting point we will write

$$\phi = [(\chi_1\gamma_1)_1(\chi_2\gamma_2)_2]$$

where the expression within the square brackets represents, as usual, the main diagonal of the determinant. The orbitals and the spin functions are denoted, respectively, by χ and γ and the subscripts outside of the parentheses label the electrons. Expansion of the determinant yields

$$\phi = (\chi_1\gamma_1)_1(\chi_2\gamma_2)_2 - (\chi_2\gamma_2)_1(\chi_1\gamma_1)_2 = (\chi_1\chi_2)(\gamma_1\gamma_2) - (\chi_2\chi_1)(\gamma_2\gamma_1)$$

$$= \phi_1\sigma_1 - \phi_2\sigma_2 \tag{10.21}$$

where ϕ_1 and ϕ_2 denote products of orbitals and σ_1 and σ_2 represent products of spin functions. In both cases the factors in the products are given according to the increasing order of the electron labels; that is

$$\phi_1 = (\chi_1)_1(\chi_2)_2 \qquad\qquad \phi_2 = (\chi_2)_1(\chi_1)_2$$

$$\sigma_1 = (\gamma_1)_1(\gamma_2)_2 \qquad\qquad \sigma_2 = (\gamma_2)_1(\gamma_1)_2$$

Equation (10.21) suggests that the correction function φ should be expressed as

$$\varphi = \varphi_1\sigma_1 - \varphi_2\sigma_2$$

so that we obtain

$$\Psi = \phi + \varphi = (\phi_1 + \varphi_1)\sigma_1 - (\phi_2 + \varphi_2)\sigma_2 = \psi_1\sigma_1 - \psi_2\sigma_2 \tag{10.22a}$$

$$\sigma_1(\mathcal{H} - E)(\phi_1 + \varphi_1) - \sigma_2(\mathcal{H} - E)(\phi_2 + \varphi_2) = 0 \qquad (10.22b)$$

$$\psi_1\sigma_1 - \psi_2\sigma_2 = \sigma_1 \frac{(\mathcal{H} - \varepsilon)(\phi_1 + \varphi_1)}{E - \varepsilon} - \sigma_2 \frac{(\mathcal{H} - \varepsilon)(\phi_2 + \varphi_2)}{E - \varepsilon} \qquad (10.22c)$$

Taking into account the linear independence of the spin function, Eq. (10.22c) may be split into

$$\psi_1 = \frac{(\mathcal{H} - \varepsilon)(\phi_1 + \varphi_1)}{E - \varepsilon} \qquad\qquad \psi_2 = \frac{(\mathcal{H} - \varepsilon)(\phi_2 + \varphi_2)}{E - \varepsilon}$$

and the computational procedure, described in the preceding sections, will be applied separately to each of these two equations. In this particular case a simplification is possible when the two spin functions are identical (Fraga and Fraga 1998a).

Multi-determinantal functions may be handled in a similar manner. We will write, for example,

$$\phi = [(\chi_1\alpha)_1(\chi_2\beta)_2] \pm [(\chi_1\beta)_1(\chi_2\alpha)_2]$$

$$= \{(\chi_1\alpha)_1(\chi_2\beta)_2 - (\chi_2\beta)_1(\chi_1\alpha)_2\} \pm \{(\chi_1\beta)_1(\chi_2\alpha)_2 - (\chi_2\alpha)_1(\chi_1\beta)_2\}$$

$$= \{(\chi_1\chi_2) \mp (\chi_2\chi_1)\}\alpha\beta - \{(\chi_2\chi_1) \pm (\chi_1\chi_2)\}\beta\alpha$$

$$= \phi_1\alpha\beta - \phi_2\beta\alpha$$

10.2.2 General Case

In order to illustrate the characteristics of the problem, for a subsequent discussion, we will consider the simple case of a closed-shell starting function ϕ for a system with four electrons and we will write

$$\phi = [(\chi_1\alpha)_1(\chi_1\beta)_2(\chi_2\alpha)_3(\chi_2\beta)_4]$$

$$= \{(\chi_1\chi_1\chi_2\chi_2) + (\chi_2\chi_2\chi_1\chi_1) - (\chi_1\chi_2\chi_2\chi_1) - (\chi_2\chi_1\chi_1\chi_2)\}(\alpha\beta\alpha\beta + \beta\alpha\beta\alpha)$$

$$+ \{(\chi_1\chi_2\chi_1\chi_2) + (\chi_2\chi_1\chi_2\chi_1) - (\chi_1\chi_1\chi_2\chi_2) - (\chi_2\chi_2\chi_1\chi_1)\}(\alpha\beta\beta\alpha + \beta\alpha\alpha\beta)$$

$$+ \{(\chi_1\chi_2\chi_2\chi_1) + (\chi_2\chi_1\chi_1\chi_2) - (\chi_1\chi_2\chi_1\chi_2) - (\chi_2\chi_1\chi_2\chi_1)\}(\alpha\alpha\beta\beta + \beta\beta\alpha\alpha)$$

$$= \phi_1(\alpha\beta\alpha\beta + \beta\alpha\beta\alpha) + \phi_2(\alpha\beta\beta\alpha + \beta\alpha\alpha\beta) + \phi_3(\alpha\alpha\beta\beta + \beta\beta\alpha\alpha)$$

$$= \phi_1\sigma_1 + \phi_2\sigma_2 + \phi_3\sigma_3 \qquad (10.23a)$$

and

$$\Psi = (\phi_1 + \varphi_1)\sigma_1 + (\phi_2 + \varphi_2)\sigma_2 + (\phi_3 + \varphi_3)\sigma_3 \qquad (10.23b)$$

For the evaluation of the energy it will suffice to apply the Schrödinger-Riccati procedure to a single function ϕ_i, at a given *PECS* (as described in Section 10.1.5), but the determination of the function Ψ requires the evaluation of each function ϕ_i, over the complete electron configuration space in each case.

10.3 Applicability of the Schrödinger-Riccati Equation

The execution of a Schrödinger-Riccati calculation involves therefore three formal steps, prior to the actual solution of the algebraic equation: determination of the function ϕ (as an eigenfunction of the total spin operators $\mathbf{S}^2$ and $\mathbf{S}_z$ and consisting of one or more determinants), expression of this function as a sum of products of spatial functions ϕ_i and spin functions σ_i, and development of the expressions for the quantities $S^{(n)}$.

The first two steps may involve a rather laborious process, particularly if the occupancy of the open shell(s) is not small. This task may be carried out, however, in a systematic way with the help of well-established techniques, thoroughly discussed in the literature. [The reader is referred to the work of Weyl (1931), Rutherford (1948), Kotani *et al.* (1955), Löwdin (1955), Kahan (1960), Kaplan (1975), Pauncz (1979), and McWeeny (1989), among others.]

The second step may be somewhat simplified if only the energy is to be evaluated. The single function ϕ_i needed may be obtained by exclusive consideration of the terms arising from the permutations in the main diagonal of the determinant(s) in ϕ, which maintain unchanged the original spin function associated with the diagonal. For example, in the case of a closed-shell function

$$\phi = [(\chi_1\alpha)_1(\chi_1\beta)_2(\chi_2\alpha)_3(\chi_2\beta)_4 \cdots (\chi_n\alpha)_{2n-1}(\chi_n\beta)_{2n}]$$

only the permutations involving separately odd-numbered and even-numbered electrons are to be considered so that all the spatial terms will be associated with the same spin function $(\alpha_1\beta_2\alpha_3\beta_4 \cdots \alpha_{2n-1}\beta_{2n})$.

The expressions for the quantities $S^{(n)}$ might be obtained using existing software packages (such as *Maple V*) for mathematical manipulations. The formulation is straightforward in the case of determinantal functions constructed from orbitals expanded in terms of Slater-type basis functions and Section 11.3 presents the expressions obtained for a two-electron system.

Finally it must be mentioned that the independence of the results for different *PECS* makes the formulation suitable for parallel computing (Hirao 1996).

11 Numerical Experience with the Schrödinger-Riccati Equation

The numerical tests of the Schrödinger-Riccati equation, presented in this chapter, will complement and illustrate with practical details the formulation and theoretical developments of the preceding chapter.

By choice, in order to obtain information on the accuracy of the calculations, the set of tests has been devised as exploratory, insofar as the correct value of the energy has been used in them.

The main body of results, obtained for the groundstate of the Hydrogen atom (Fraga and Fraga 1998b), has been complemented with some additional results for a simple one-dimensional Schrödinger equation. These results are presented and discussed below but first the nature of the starting functions ϕ and the precision difficulties in the calculations merit some general comments.

The starting functions used in both cases are of the form of the exact solution and therefore they can evolve into the latter when the appropriate value of the variational parameter is used. They will constitute, however, poor starting functions if a poor choice of the parameter is made.

Preliminary calculations carried out for the groundstate of the Helium atom (Fraga and Fraga 1998b), using the correlated closed- and open-shell functions of Roothaan and Weiss (1960), have served to highlight the precision problems of a predictive calculation. The energies predicted at different $PECS$ showed a great sensitivity to small changes in the values of the correction function φ. That is, the direct evaluation of the energy, using the value of φ obtained from the algebraic Schrödinger-Riccati equation, yielded a very poor value but a small subsequent change in the value of φ was sufficient for a correct prediction. Taking into account the high quality of the starting function, this behaviour may reflect the double dependence, existing in a predictive calculation, on the number of terms of the expansions used in the calculations. First, the value obtained for φ will depend on the degree of the algebraic Schrödinger-Riccati equation, Eq. (10.15). Then, the value predicted for the energy will depend on both the value of φ and on the number of terms in the expansions in equations of the type of Eqs. (10.20). Therefore it should be emphasized that the comments made below regarding what could be the appropriate degree of algebraic equation to be used should be taken simply as indicative of its general range, with the expectation that a higher number of terms in all the expansions may probably be needed.

The manner in which the numerical calculations are performed may also have a considerable effect on the precision of the results and this will be examined in Section 11.3, where the formulation is applied to the groundstates of the ions of the *He*-isoelectronic series.

11.1 1-Dimensional Schrödinger Equation

The function

$$
\Psi(x) = \begin{cases} 0 & \text{for } x < 0 \\ xe^{-x} & \text{for } 0 \leq x \end{cases}
$$

is an eigenfunction of the one-dimensional Schrödinger equation (in atomic units)

$$
-\frac{1}{2}\frac{d^2\Psi(x)}{dx^2} - \frac{1}{x}\Psi(x) = E\Psi(x)
$$

with an eigenvalue $E = -0.5$ hartrees.

The function $\phi = xe^{-\zeta x}$, which may evolve (for $\zeta = 1$) into the correct eigenfunction, has been used in the calculations. The quantities $H^{(n)}$ [see Eqs. (10.3)] are given in this case by

$$
H^{(0)} \equiv \mathcal{H}\phi = \{ -\frac{1}{2}\zeta^2 x + (\zeta - 1) \} e^{-\zeta x} \tag{11.1a}
$$

$$
H^{(1)} = -\frac{1}{x^2} + \frac{2\zeta - 1}{x} - \frac{1}{2}\zeta^2 \tag{11.1b}
$$

$$
H^{(n)} = (-1)^n \{ a_n + b_n \zeta x \} x^{-(n+1)} e^{(n-1)\zeta x} \quad \text{(for } n > 1) \tag{11.1c}
$$

with

$$
a_2 = -2 \qquad b_2 = 1 \qquad a_{n+1} = (n-2)a_n + b_n \qquad b_{n+1} = (n-2)b_n
$$

For simplicity in the presentation of the results we define the quantities

$$t_n = \frac{1}{n!} H^{(n)} \varphi^n$$

so that the Schrödinger-Riccati equation, Eq. (10.13), becomes

$$\sum_{n=0}^{t} \frac{1}{n!} S^{(n)} \varphi^n = (H^{(0)} - E\phi) + (H^{(1)} - E)\varphi + \sum_{n=2}^{t} \frac{1}{n!} H^{(n)} \varphi^n = -E\Psi + \sum_{n=0}^{t} t_n = 0$$

when using the correct value of φ.

The values of $H^{(n)}$ and of the terms t_n, obtained at $x = 1$ bohr with $\zeta = 0.25$ and using the correct value of φ, are presented in Table 11.1. They illustrate clearly the precision problems that may arise in the solution of a high-degree Schrödinger-Riccati equation.

Table 11.1. Values[a] of $H^{(n)}$ and t_n, at $x = 1$ bohr, for the one-dimensional Schrödinger equation, obtained with a function $\phi = xe^{-\zeta x}$, $\zeta = 0.25$.

n	$H^{(n)}$	t_n	n	$H^{(n)}$	t_n
0	-0.60843811(0)	-0.60843811	8	0.29003197(4)	0.00005848
1	-0.15312500(1)	0.62922330	9	-0.31388710(5)	0.00002890
2	-0.22470445(1)	-0.18971386	10	0.37024940(6)	0.00001401
3	0.12365410(1)	-0.01429991	11	-0.47698850(7)	0.0000674
4	-0.10585000(1)	-0.00125752	12	0.66922933(8)	0.00000324
5	-0.13591409(1)	0.00013270	13	-0.10181246(10)	0.00000156
6	0.27922744(2)	0.00018671	14	0.16717042(11)	0.00000075
7	-0.28682810(3)	0.00011259	15	-0.29490875(12)	0.00000036

[a]See the text for details. The values of t_n have been calculated with the correct value of φ. The numbers in parentheses denote powers of 10.

Table 11.2 collects the values of the energy obtained using the above equation, at $x = 1$ bohr and with $\zeta = 0.25$, 0.50, and 0.90, with the correct values of φ. These results indicate that in a predictive calculation, using the correct value of the energy, the minimum degree of the Schrödinger-Riccati equation to be solved in order to obtain an accurate value of φ (to the precision considered in the tests) could be 13, 9, and 3, respectively, for the three cases considered. That is: the necessary degree of the Schrödinger-Riccati equation decreases as the quality of the function ϕ improves.

Table 11.2. Values[a] of E (hartrees) for the one-dimensional Schrödinger equation, obtained at $x = 1$ bohr from a Schrödinger-Riccati equation of degree n with a function $\phi = xe^{-\zeta x}$.

n	$\zeta = 0.25$	$\zeta = 0.50$	$\zeta = 0.90$
0	-0.78125[b]	-0.62500[b]	0.50500[b]
1	0.05650	-0.30064	-0.49448
2	-0.45920	-0.49208	-0.49999
3	-0.49807	-0.50045	**-0.50000**
4	-0.50149	-0.50045	
5	-0.50112	-0.50019	
6	-0.50062	-0.50007	
7	-0.50031	-0.50002	
8	-0.50015	-0.50001	
9	-0.50007	**-0.50000**	
10	-0.50004		
11	-0.50002		
12	-0.50001		
13	**-0.50000**		

[a]Using the correct values of φ.
[b]These are the values of ε_ϕ [see Eqs. (10.6)].

It is interesting to note that the error in the local energy ε_ϕ for $\zeta = 0.90$ is of the order of 1% of the exact energy. Taking into account that, as a rule, the average correlation energy for a Hartree-Fock function is of the order of 1% of the exact energy (Frye *et al.* 1990), extrapolation of the present results would suggest that a Schrödinger-Riccati equation of a low degree could be appropriate when using a starting function ϕ of Hartree-Fock quality, at least for some *PECS*.

11.2 The Groundstate of the Hydrogen Atom

The Hamiltonian operator (in polar coordinates and atomic units, with a reduced mass $\mu = 1$) for the groundstate of the Hydrogen atom is

$$\mathcal{H} = -\frac{1}{2r^2}\frac{d}{dr}(r^2\frac{d}{dr}) - \frac{1}{r}$$

The groundstate function is $\Psi = (1/\sqrt{\pi})\, e^{-r}$ and all the calculations in this section have been performed with an unnormalized function of the form $\phi = (1/\sqrt{\pi})\, e^{-\zeta r}$, which is equivalent to the exact function at $\zeta = 1$.

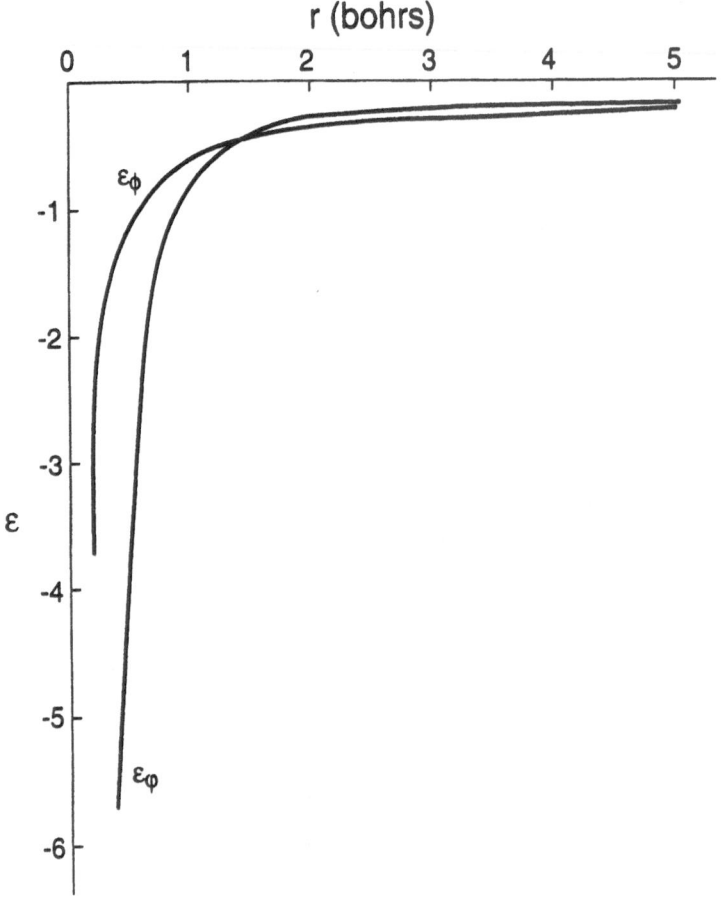

Figure 11.1. Behaviour of ε_ϕ and ε_φ, as functions of r, for the 1s-orbital of the Hydrogen atom, obtained with $\phi = (1/\sqrt{\pi}\,)e^{-\zeta r}$, with $\zeta = 0.5$.

For illustrative purposes, Figure 11.1 presents an example (for $\zeta = 0.5$) of the behaviour of both ε_ϕ and ε_φ as functions of r.

11.2.1 Preliminary Test

We will consider first a test, equivalent to that presented in the preceding section, but with a different approach. The quantities $S^{(n)}$ [see Eqs. (10.3)], for $n > 0$, may be expressed as

$$S^{(n)} = s^{(n)} \phi^{-(n-1)}$$

where the quantities $s^{(n)}$ depend on r and ζ [see, e.g., Eqs. (11.1)]. Therefore, the Schrödinger-Riccati equation, Eq. (10.13), may be written as

$$\sum_{n=0}^{t} \frac{1}{n!} S^{(n)} \varphi^n = S^{(0)} + \sum_{n=1}^{t} \frac{1}{n!} s^{(n)} \phi^{-(n-1)} \varphi^n$$

$$= S^{(0)} + \varphi \sum_{n=1}^{t} \frac{1}{n!} s^{(n)} \phi^{-(n-1)} \varphi^{n-1} = 0 \tag{11.2}$$

Table 11.3. Values[a] of $|\Delta w|$ for the $1s$-orbital of the Hydrogen atom, obtained with a function $\phi = (1/\sqrt{\pi})e^{-\zeta r}$.

r (bohrs)	$\zeta = 0.25$	$\zeta = 0.50$	$\zeta = 0.75$
0.2	0.0	0.0	0.0
0.4	0.0	0.0	0.0
0.6	0.00001	0.0	0.0
0.8	0.00004	0.0	0.0
1.0	0.00022	0.00003	0.0
2.0	0.01259	0.00124	0.00002
3.0	0.01675	0.00419	0.00011
4.0	0.02085	0.00885	0.00048
5.0	0.02183	0.01362	0.00131

[a]The exact values of w were used in the *rhs* of Eq. (11.3), with a nine-term expansion.

Table 11.4. Values[a,b] (in hartrees) of ε_φ for the 1s-orbital of the Hydrogen atom obtained with a function $\phi = (1/\sqrt{\pi}\,)e^{-\zeta r}$.

r (bohrs)	quadratic	theoretical cubic	quartic	exact
$\zeta = 0.25$[c]				
0.2	-24.14931	-24.05966	-24.05668	-24.05663
0.4	-6.01813	-5.93100	-5.92577	-5.92573
0.6	-2.75032	-2.66294	-2.65555	-2.65593
0.8	-1.64106	-1.54826	-1.53671	-1.53892
1.2	-0.71665	-0.76286	-0.76726	-0.76330
2.0	-0.38212	-0.37668	-0.37668	-0.37932
2.5	-0.29893	-0.29429	-0.29490	-0.30068
3.0	-0.24860	-0.24583	-0.24702	-0.25548
4.0	-0.19036	-0.19036	-0.19218	-0.20401
5.0	-0.15730	-0.15869	-0.16070	-0.17357
$\zeta = 0.50$				
0.2	-22.86879	-22.83101	-22.83021	-22.83021
0.4	-5.36287	-5.32824	-5.32704	-5.32707
0.6	-2.30235	-2.26944	-2.26816	-2.26839
0.8	-1.29385	-1.25857	-1.25743	-1.25831
1.2	-0.58722	-0.59342	-0.59308	-0.59235
1.6	-0.38653	-0.38472	-0.38545	-0.38650
2.0	-0.29775	-0.29775	-0.29957	-0.30225

167

Table 11.4.   Continued

r		theoretical		exact
(bohrs)	quadratic	cubic	quartic	
2.5	-0.24399	-0.24673	-0.24972	-0.25473
3.0	-0.21575	-0.22055	-0.22440	-0.23183
4.0	-0.18750	-0.19447	-0.19918	-0.21087
5.0	-0.17342	-0.18106	-0.18589	-0.20041
$\zeta = 0.75$				
0.2	-21.65391	-21.64501	-21.64492	-21.64492
0.4	-4.77660	-4.76911	-4.76901	-4.76901
0.6	-1.92715	-1.92089	-1.92085	-1.92088
0.8	-1.02250	-1.01698	-1.01711	-1.01719
1.2	-0.45987	-0.45965	-0.45977	-0.45981
1.6	-0.30840	-0.30951	-0.31016	-0.31042
2.0	-0.25712	-0.26001	-0.26119	-0.26173
2.5	-0.23674	-0.24158	-0.24344	-0.24448
3.0	-0.23273	-0.23914	-0.24167	-0.24335
4.0	-0.23734	-0.24583	-0.24951	-0.25282
5.0	-0.24435	-0.25386	-0.25831	-0.26349

[a]Using the correct value E = -0.5 hartrees.
[b]Reprinted from the work of Fraga and Fraga (1998b) with permission of the Polish Academy of Sciences.
[c]At r = 1.6 bohrs, $S^{(0)} = 0$.

Taking into account [see Eqs. (10.6)] that

$$S^{(0)} = \mathcal{H}\phi - E\phi = (\varepsilon_\phi - E)\phi$$

and defining the quantity $w = \phi/\phi$, Eq. (11.2) may be rewritten (with $\phi \neq 0$) as

$$(\varepsilon_\phi - E)\phi + \phi \sum_{n=1}^{t} \frac{1}{n!} s^{(n)} w^{n-1} = \{(\varepsilon_\phi - E) + w \sum_{n=1}^{t} \frac{1}{n!} s^{(n)} w^{n-1}\}\phi = 0$$

and finally as

$$w = \frac{E - \varepsilon_\phi}{\displaystyle\sum_{n=1}^{t} \frac{1}{n!} s^{(n)} w^{n-1}} \tag{11.3}$$

This equation may be used for a test as follows. Using the correct values of E and ϕ, both known in the present case, in the *rhs* of the equation one can evaluate a new value of w. The difference, $|\Delta w|$, between the input and predicted values of w, will illustrate the dependence on the quality of the function ϕ. The results presented in Table 11.3 confirm that the accuracy will improve with a better quality of ϕ and that the prediction of the tails of the functions will face more precision difficulties.

11.2.2 Evaluation of ϕ and ε_ϕ

The calculations for the evaluation of the local Riccati energy, ε_ϕ [see Eqs. (10.6)], have been performed, using the correct value of E, with the quadratic, cubic, and quartic approximations of the Schrödinger-Riccati equation (see Section 10.1.4). Table 11.4 presents the values of the roots of those equations which are closest to the correct value. These results show again the dependence on the quality of the function ϕ (as determined by the value of ζ) and on the degree of the Schrödinger-Riccati equation as well as the worsening of the accuracy as the separation from the nucleus increases.

The direct evaluation of the values of the correction function ϕ, using again the correct value of E, has also been carried out for completeness and the results are presented graphically in Figs. 11.2-11.3, which confirm once more the above conclusions. These figures have been prepared using the real roots of the quadratic, cubic, and quartic Schrödinger-Riccati equations which are closest to the correct value of ϕ. A straightforward and excellent prediction is obtained, for $\zeta = 0.5$, from the solutions of both the cubic and the quartic equations (Fig. 11.2), while it is interesting to observe the necessary cross-over from one root to another when using the solutions of the quadratic equation (Fig. 11.3).

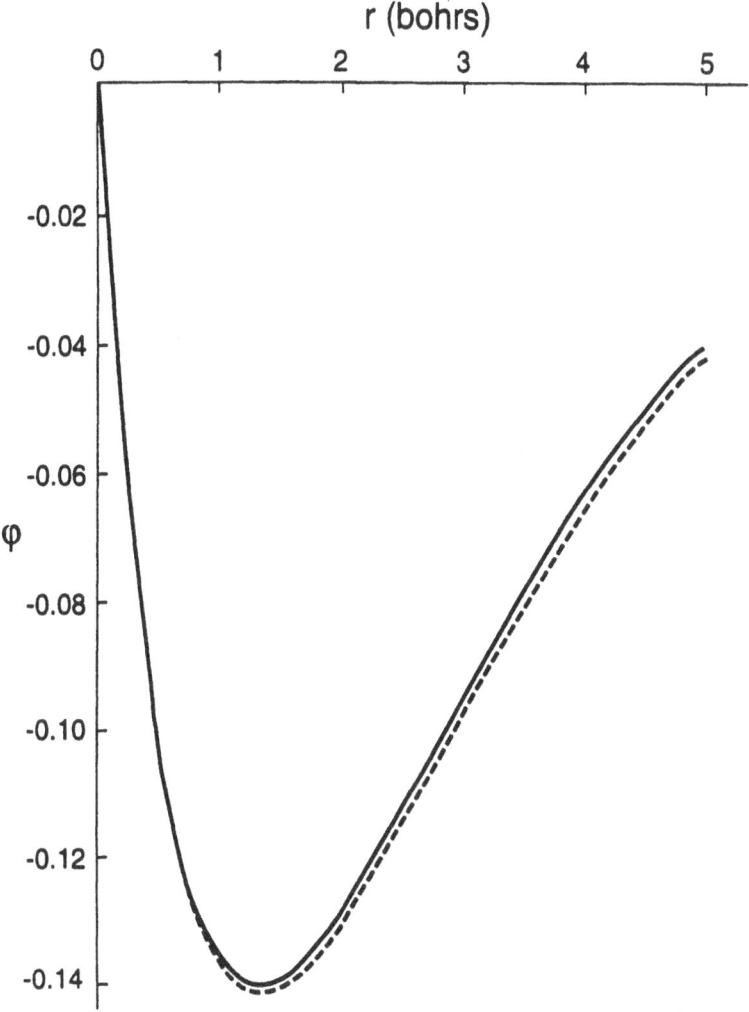

Figure 11.2. Behaviour of φ for the $1s$-orbital of the Hydrogen atom. The full line corresponds to the cubic and quartic predictions (which cannot be distinguished for the scale used) obtained with $\phi = (1/\sqrt{\pi}\,)e^{-\zeta r}$, with $\zeta = 0.5$. The dashed line corresponds to the exact solution.

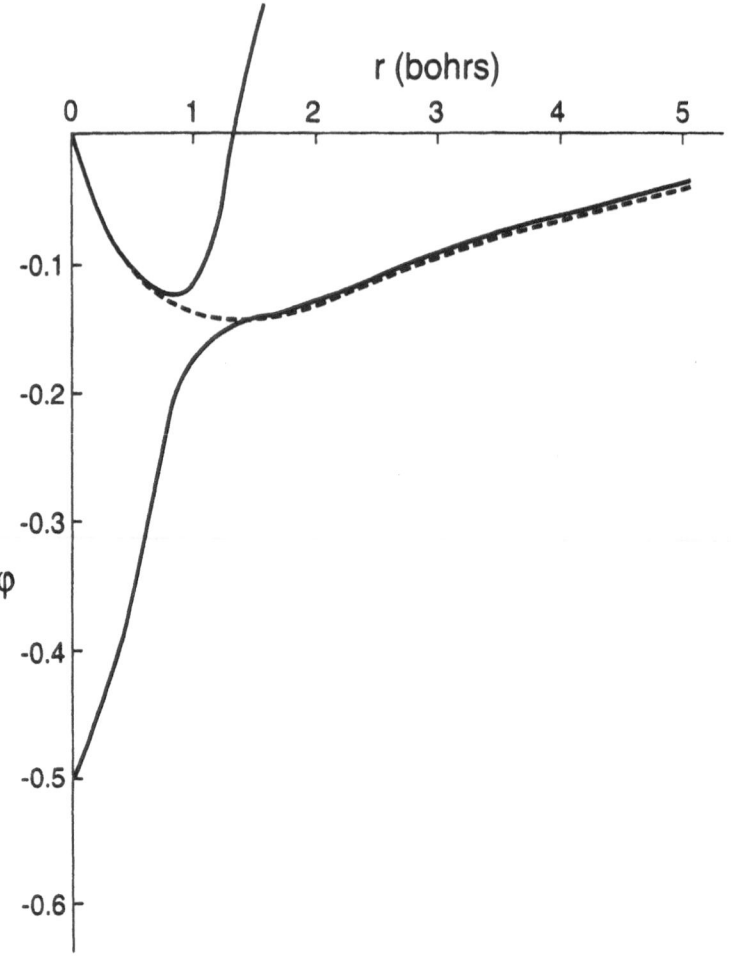

Figure 11.3. Behaviour of φ for the 1*s*-orbital of the Hydrogen atom. The full lines represent the two roots of the quadratic equation, obtained with $\phi = (1/\sqrt{\pi}\,)e^{-\zeta r}$, with $\zeta = 0.5$. The dashed line corresponds to the exact solution.

11.3 Groundstates of Ions of the He-Isoelectronic Series

Functions of a high quality, available for a variety of systems, may be used as starting functions in a Schrödinger-Riccati calculation but it may be necessary to implement the corresponding formulation independently for each separate case (depending on the characteristics of the function). On the other hand, Hartree-Fock functions (or other determinantal functions of similar quality) are available or may be easily obtained for many systems. For example, analytical Hartree-Fock functions for atoms may be found in existing compilations [such as that of Clementi (1965)] or may be generated from the numerical functions obtained with existing software (Froese Fischer 1977).

It is for this reason that in this section we present, as an example, the formulation for all the quantities needed in the application of the Schrödinger-Riccati equation to determinantal functions. The groundstate of a two-electron atomic system (*i.e.*, a member of the *He*-isoelectronic series) has been chosen for simplicity. This presentation includes suggestions, based on preliminary numerical observations, on how to overcome possible precision difficulties to be encountered.

The Hamiltonian operator for the groundstates of the systems under consideration will be written (Roothaan and Weiss 1960) as

$$\mathcal{H} = \sum_{i=1}^{2} \mathcal{T}_i + \mathcal{T}_{12} + \mathcal{T}_{12}' + \mathcal{V}$$

where

$$\mathcal{T}_i = -\frac{1}{2}\left(\frac{\partial^2}{\partial r_i^2} + \frac{2}{r_i}\frac{\partial}{\partial r_i}\right)$$

$$\mathcal{T}_{12} = -\left(\frac{\partial^2}{\partial r_{12}^2} + \frac{2}{r_{12}}\frac{\partial}{\partial r_{12}}\right)$$

$$\mathcal{T}_{12}' = -\frac{1}{2}\left(\cos\theta_1 \frac{\partial}{\partial r_1} + \cos\theta_2 \frac{\partial}{\partial r_2}\right)\frac{\partial}{\partial r_{12}}$$

are the kinetic energy operators and

$$\mathcal{V} = -\frac{Z}{r_1} - \frac{Z}{r_2} + \frac{1}{r_{12}}$$

is the potential energy operator. Z denotes the nuclear charge and θ_1 and θ_2 are the angles formed by the vectors $\mathbf{r}_1$ and $\mathbf{r}_{12}$ and $\mathbf{r}_2$ and $\mathbf{r}_{21}$, respectively.

The starting function for an energy calculation (see Section 10.2) is

$$\phi = \chi(1)\chi(2)$$

(where the numbers in parentheses label the electrons), with the orbital χ expanded

$$\chi = \sum_k \xi_k c_k$$

in terms of $1s$ Slater-type basis functions

$$\xi_k = e^{-\zeta_k r}$$

The expansion coefficients c_k and the orbital exponents ζ_k are available in the work of Clementi (1965).

All the terms in the Hamiltonian operator will be used in the evaluation of the contribution of the correction function φ to the energy. On the other hand, the quantities $S^{(n)}$, appearing in the Schrödinger-Riccati equation, will be evaluated using only the terms $\mathcal{T}_1$, $\mathcal{T}_2$, and $\mathcal{V}$ because the function ϕ is independent of r_{12}. Taking into account that $\mathcal{V}$ depends on r_1, r_2, and r_{12} it will be more convenient, in the present case, to write

$$S^{(0)} = (\mathcal{T}_1 + \mathcal{T}_2 + \mathcal{V} - E)\phi = (H^{(0)} - E\phi) = T^{(0)} + (\mathcal{V} - E)\phi$$

$$S^{(1)} = T^{(1)} + (\mathcal{V} - E) \qquad\qquad S^{(n)} = T^{(n)} \quad (n > 1)$$

with

$$T^{(n)} = \frac{\partial^n \mathcal{T}\phi}{\partial \varphi^n} = \frac{\partial^n (\mathcal{T}_1 + \mathcal{T}_2)\phi}{\partial \varphi^n}$$

so that the Schrödinger-Riccati equation becomes

$$(\mathcal{V} - E)\phi + (\mathcal{V} - E)\varphi + \sum_{n=0} \frac{1}{n!} T^{(n)} \varphi^n = 0 \tag{11.4}$$

The evaluation of the quantities $T^{(n)}$ is described below. In order to simplify the notation we will denote r_{12} as R and use the notation ∂_r and ∂_r^2 to indicate differentiation with respect to r (either r_1 or r_2) and ∂_R and ∂_R^2 to indicate differentiation with respect to R. Whenever the label of the electron must be included, the above notation for differentiation with respect to r will be further simplified to ∂_1, ∂_1^2, ∂_2, and ∂_2^2, respectively, and the successive differentiation with respect to r (either r_1 or r_2) and R will be represented by ∂_{1R} and ∂_{2R}, respectively. In addition, whenever it is necessary to indicate that differentiation of a quantity depending on both electrons has been carried out for electron i, this dependence will be indicated by the corresponding subindex; thus, for example, the notation $T^{(n)}_{(1;i)}$ and $T^{(n)}_{(2;i)}$ indicates that differentiation of the quantity $T^{(n)}$ has been carried out with respect to electron i, i.e., $T^{(n)}_{(1;i)} = \partial_{r_i} T^{(n)}$ and $T^{(n)}_{(2;i)} = \partial^2_{r_i} T^{(n)}$. The dependence of any of the quantities to be evaluated on the electron label will be indicated by the corresponding label in parentheses; for example, $\xi_k(i)$ will stand for $\exp(-\zeta_k r_i)$. Any of the orbital exponents may be chosen as variational parameter (in the sense described in Section 10.1.1) and it will be denoted as ζ.

For an extension of the formulation presented below to many-electron systems with basis functions given by a polynomial times an exponential, the equations in Section 11.3.1 must be changed in order to introduce the contribution from the polynomial. The equations, in Section 11.3.2, for the quantities $\phi^{(n)}$ and $t^{(n)}$ (as well as for their derivatives) should be changed in order to take into account the increased number of orbitals in the monomial term as well as for the possible increase in the number of monomial terms. The formulation for the quantities $T^{(n)}$ (and their derivatives), however, is general.

11.3.1 Evaluation of the Basic Quantities

The basic quantities needed for the calculations are:

$$\xi^{(0)}_k \equiv \xi_k$$

$$\xi^{(n)}_k = \frac{\partial^n \xi^{(0)}_k}{\partial \zeta^n} = \begin{cases} (-1)^n r^n \xi^{(0)}_k & (\zeta_k = \zeta) \\ 0 & (\zeta_k \neq \zeta) \end{cases}$$

$$\xi^{(0)}_{k(1)} = \partial_r \xi^{(0)}_k = -\zeta_k \xi^{(0)}_k$$

$$\xi^{(n)}_{k(1)} = \partial_r \xi^{(n)}_k = \frac{\partial^n \xi^{(0)}_{k(1)}}{\partial \zeta^n} = \begin{cases} (-1)^n (n r^{n-1} - \zeta_k r^n) \xi^{(0)}_k & (\zeta_k = \zeta) \\ 0 & (\zeta_k \neq \zeta) \end{cases}$$

$$\xi^{(0)}_{k(2)} = \partial^2_r \xi^{(0)}_k = \zeta^2_k \xi^{(0)}_k$$

$$\xi_{k(2)}^{(n)} = \partial_r^2 \xi_k^{(n)} = \frac{\partial^n \zeta_{k(2)}^{(0)}}{\partial \zeta^n} = \begin{cases} (-1)^n [n(n-1)r^{n-2} - 2n\zeta_k r^{n-1} + \zeta_k^2 r^n] \xi_k^{(0)} & (\zeta_k = \zeta) \\ \\ 0 & (\zeta_k \neq \zeta) \end{cases}$$

$$\chi^{(0)} \equiv \chi \qquad\qquad\qquad \chi^{(n)} = \sum_k \xi_k^{(n)} c_k$$

$$\chi_{(1)}^{(0)} = \sum_k \xi_{k(1)}^{(0)} c_k \qquad\qquad \chi_{(1)}^{(n)} = \sum_k \xi_{k(1)}^{(n)} c_k$$

$$\chi_{(2)}^{(0)} = \sum_k \xi_{k(2)}^{(0)} c_k \qquad\qquad \chi_{(2)}^{(n)} = \sum_k \xi_{k(2)}^{(n)} c_k$$

$$t_k^{(0)}(i) = \mathcal{T}_i \xi_k^{(0)}(i) = U_k^{(0)}(i) \xi_k^{(0)}(i)$$

$$\text{with } U_k^{(0)}(i) = \frac{1}{2} \left(\frac{2\zeta_k}{r_i} - \zeta_k^2 \right)$$

$$t_k^{(1)}(i) = \frac{\partial t_k^{(0)}(i)}{\partial \zeta} = \begin{cases} U_k^{(1)}(i) \xi_k^{(0)}(i) & (\zeta_k = \zeta) \\ \\ 0 & (\zeta_k \neq \zeta) \end{cases}$$

$$\text{with } U_k^{(1)}(i) = \frac{1}{2} \left(\frac{2}{r_i} - 4\zeta_k + \zeta_k^2 r_i \right)$$

$$t_k^{(2)}(i) = \frac{\partial^2 t_k^{(0)}(i)}{\partial \zeta^2} = \begin{cases} U_k^{(2)}(i) \xi_k^{(0)}(i) & (\zeta_k = \zeta) \\ \\ 0 & (\zeta_k \neq \zeta) \end{cases}$$

$$\text{with } U_k^{(2)}(i) = \frac{1}{2} \left(-6 + 6\zeta_k r_i - \zeta_k^2 r_i^2 \right)$$

$$t_k^{(n)}(i) = \frac{\partial^n t_k^{(0)}(i)}{\partial \zeta^n} = \begin{cases} U_k^{(n)}(i)\xi_k^{(0)}(i) & (\zeta_k = \zeta) \\ \\ 0 & (\zeta_k \neq \zeta) \end{cases}$$

$$\text{with } U_k^{(n)}(i) = \frac{1}{2}(a_{1n}r_i^{n-2} + a_{2n}\zeta_k r_i^{n-1} + a_{3n}\zeta_k^2 r_i^n)$$

$$a_{1n} = (-1)^{n+1} n(n+1) \qquad a_{2n} = (-1)^n 2(n+1) \qquad a_{3n} = (-1)^{n+1}$$

$$t_{k(1)}^{(0)}(i) = \partial_i t_k^{(0)}(i) = V_{k(1)}^{(0)}(i)\xi_k^{(0)}(i)$$

$$\text{with } V_{k(1)}^{(0)}(i) = U_{k(1)}^{(0)}(i) - \zeta_k U_k^{(0)}(i)$$

$$U_{k(1)}^{(0)}(i) = -\frac{\zeta_k}{r_i^2}$$

$$t_{k(1)}^{(1)}(i) = \partial_i t_k^{(1)}(i) = \begin{cases} V_{k(1)}^{(1)}(i)\xi_k^{(0)}(i) & (\zeta_k = \zeta) \\ \\ 0 & (\zeta_k \neq \zeta) \end{cases}$$

$$\text{with } V_{k(1)}^{(1)} = U_{k(1)}^{(1)}(i) - \zeta_k U_k^{(1)}(i)$$

$$U_{k(1)}^{(1)}(i) = \frac{1}{2}(-\frac{2}{r_i^2} + \zeta_k^2)$$

$$t_{k(1)}^{(2)}(i) = \partial_i t_k^{(2)}(i) = \begin{cases} V_{k(1)}^{(2)}(i)\xi_k^{(0)} & (\zeta_k = \zeta) \\ \\ 0 & (\zeta_k \neq \zeta) \end{cases}$$

$$\text{with } V_{k(1)}^{(2)}(i) = U_{k(1)}^{(2)}(i) - \zeta_k U_k^{(2)}(i)$$

$$U_{k(1)}^{(2)}(i) = \frac{1}{2} \left(6\zeta_k - 2\zeta_k^2 r_i \right)$$

$$t_{k(1)}^{(n)}(i) = \partial_i t_k^{(n)}(i) = \begin{cases} V_{k(1)}^{(n)}(i)\xi_k^{(0)}(i) & (\zeta_k = \zeta) \\ \\ 0 & (\zeta_k \neq \zeta) \end{cases}$$

with $V_{k(1)}^{(n)}(i) = U_{k(1)}^{(n)}(i) - \zeta_k U_k^{(n)}(i)$

$$U_{k(1)}^{(n)}(i) = \frac{1}{2} \left[a_{1n}(n-2) r_i^{n-3} + a_{2n}(n-1)\zeta_k r_i^{n-2} + a_{3n}n\zeta_k^2 r_i^{n-1} \right]$$

$$t_{k(2)}^{(0)}(i) = \partial_i t_{k(1)}^{(0)}(i) = W_{k(2)}^{(0)}\zeta_k^{(0)}(i)$$

with $W_{k(2)}^{(0)}(i) = V_{k(2)}^{(0)}(i) - \zeta_k V_{k(1)}^{(0)}(i)$

$$V_{k(2)}^{(0)}(i) = U_{k(2)}^{(0)}(i) - \zeta_k U_{k(1)}^{(0)}(i)$$

$$U_{k(2)}^{(0)}(i) = \frac{2\zeta_k}{r_i^3}$$

$$t_{k(2)}^{(1)}(i) = \partial_i t_{k(1)}^{(1)}(i) = \begin{cases} W_{k(2)}^{(1)}(i)\xi_k^{(0)}(i) & (\zeta_k = \zeta) \\ \\ 0 & (\zeta_k \neq \zeta) \end{cases}$$

with $W_{k(2)}^{(1)}(i) = V_{k(2)}^{(1)}(i) - \zeta_k V_{k(1)}^{(1)}$

$$V_{k(2)}^{(1)}(i) = U_{k(2)}^{(1)}(i) - \zeta_k U_{k(1)}^{(1)}(i)$$

$$U_{k(2)}^{(1)}(i) = \frac{2}{r_i^3}$$

$$t_{k(2)}^{(2)}(i) = \partial_i t_{k(1)}^{(2)}(i) = \begin{cases} W_{k(2)}^{(2)}(i)\xi_k^{(0)}(i) & (\zeta_k = \zeta) \\ \\ 0 & (\zeta_k \neq \zeta) \end{cases}$$

with $W_{k(2)}^{(2)}(i) = V_{k(2)}^{(2)}(i) - \zeta_k V_{k(1)}^{(2)}(i)$

$$V_{k(2)}^{(2)}(i) = U_{k(2)}^{(2)}(i) - \zeta_k U_{k(1)}^{(2)}(i)$$

$$U_{k(2)}^{(2)}(i) = -\zeta_k^2$$

$$t_{k(2)}^{(n)}(i) = \partial_i t_{k(1)}^{(n)}(i) = \begin{cases} W_{k(2)}^{(n)}(i)\xi_k^{(0)}(i) & (\zeta_k = \zeta) \\ \\ 0 & (\zeta_k \neq \zeta) \end{cases}$$

with $W_{k(2)}^{(n)}(i) = V_{k(2)}^{(n)}(i) - \zeta_k V_{k(1)}^{(n)}(i)$

$$V_{k(2)}^{(n)}(i) = U_{k(2)}^{(n)}(i) - \zeta_k U_{k(1)}^{(n)}(i)$$

$$U_{k(2)}^{(n)}(i) = \frac{1}{2}[\, a_{1n}(n-2)(n-3)\, r_i^{n-4}$$

$$+ a_{2n}(n-1)(n-2)\zeta_k r_i^{n-3} + a_{3n}n(n-1)\zeta_k^2\, r_i^{n-2}]$$

$$t^{(n)}(i) = \sum_k t_k^{(n)}(i)c_k$$

$$t_{(j;i)}^{(n)}(i) = \sum_k t_{k(j;i)}^{(n)}(i)c_k \qquad (j = 1,2;\, i = 1,2)$$

11.3.2 Evaluation of the Quantities $T^{(n)}$

We first define the quantities

$$\phi^{(0)} \equiv \phi = \chi(1)\chi(2)$$

$$\phi^{(n)} = \frac{\partial^n \phi^{(0)}}{\partial \zeta^n} = \sum_{m=0}^{n} \frac{n!}{m!(n-m)!} \left[\frac{\partial^{n-m} \chi(1)}{\partial \zeta^{n-m}} \right] \left[\frac{\partial^m \chi(2)}{\partial \zeta^m} \right]$$

$$= \sum_{m=0}^{n} \frac{n!}{m!(n-m)!} \chi^{(n-m)}(1) \chi^{(m)}(2)$$

$$\phi^{(0)}_{(1;1)} = \partial_1 \phi^{(0)} = \chi^{(0)}_{(1;1)}(1) \chi^{(0)}(2)$$

$$\phi^{(0)}_{(1;2)} = \partial_2 \phi^{(0)} = \chi^{(0)}(1) \chi^{(0)}_{(1;2)}(2)$$

$$\phi^{(n)}_{(1;1)} = \partial_1 \phi^{(n)} = \sum_{m=0}^{n} \frac{n!}{m!(n-m)!} \chi^{(n-m)}_{(1;1)}(1) \chi^{(m)}(2)$$

$$\phi^{(n)}_{(1;2)} = \partial_2 \phi^{(n)} = \sum_{m=0}^{n} \frac{n!}{m!(n-m)!} \chi^{(n-m)}(1) \chi^{(m)}_{(1;2)}(2)$$

$$\phi^{(0)}_{(2;1)} = \partial_1^2 \phi^{(0)} = \chi^{(0)}_{(2;1)}(1) \chi^{(0)}(2)$$

$$\phi^{(0)}_{(2;2)} = \partial_2^2 \phi^{(0)} = \chi^{(0)}(1) \chi^{(0)}_{(2;2)}(2)$$

$$\phi^{(n)}_{(2;1)} = \partial_1^2 \phi^{(n)} = \sum_{m=0}^{n} \frac{n!}{m!(n-m)!} \chi^{(n-m)}_{(2;1)}(1) \chi^{(m)}(2)$$

$$\phi^{(n)}_{(2;2)} = \partial_2^2 \phi^{(n)} = \sum_{m=0}^{n} \frac{n!}{m!(n-m)!} \chi^{(n-m)}(1) \chi^{(m)}_{(2;2)}(2)$$

as well as

$$t^{(0)} = t^{(0)}(1) \chi^{(0)}(2) + \chi^{(0)}(1) t^{(0)}(2)$$

$$t^{(n)} = \frac{\partial^n t^{(0)}}{\partial \zeta^n} = \sum_{m=0}^{n} \frac{n!}{m!(n-m)!} \left\{ \left[\frac{\partial^{n-m} t^{(0)}(1)}{\partial \zeta^{n-m}} \right] \left[\frac{\partial^m \chi(2)}{\partial \zeta^m} \right] + \left[\frac{\partial^{n-m} \chi(1)}{\partial \zeta^{n-m}} \right] \left[\frac{\partial^m t^{(0)}(2)}{\partial \zeta^m} \right] \right\}$$

$$= \sum_{m=0}^{n} \frac{n!}{m!(n-m)!} \{ t^{(n-m)}(1)\chi(2) + \chi(1)t^{(m)}(2) \}$$

$$t_{(1;1)}^{(0)} = \partial_1 t^{(0)} = t_{(1;1)}^{(0)}(1)\chi^{(0)}(2) + \chi_{(1;1)}^{(0)}(1)t^{(0)}(2)$$

$$t_{(1;2)}^{(0)} = \partial_2 t^{(0)} = t^{(0)}(1)\chi_{(1;2)}^{(0)}(2) + \chi^{(0)}(1)t_{(1;2)}^{(0)}(2)$$

$$t_{(2;1)}^{(0)} = \partial_1^2 t^{(0)} = t_{(2;1)}^{(0)}(1)\chi^{(0)}(2) + \chi_{(2;1)}^{(0)}(1)t^{(0)}(2)$$

$$t_{(2;2)}^{(0)} = \partial_2^2 t^{(0)} = t^{(0)}(1)\chi_{(2;2)}^{(0)}(2) + \chi^{(0)}(1)t_{(2;2)}^{(0)}(2)$$

$$t_{(1;1)}^{(n)} = \partial_1 t^{(n)} = \sum_{m=0}^{n} \frac{n!}{m!(n-m)!} \{ t_{(1;1)}^{(n-m)}(1)\chi(2) + \chi_{(1;1)}(1)t^{(m)}(2) \}$$

$$t_{(1;2)}^{(n)} = \partial_2 t^{(n)} = \sum_{m=0}^{n} \frac{n!}{m!(n-m)!} \{ t^{(n-m)}(1)\chi_{(1;2)}(2) + \chi(1)t_{(1;2)}^{(m)}(2) \}$$

$$t_{(2;1)}^{(n)} = \partial_1^2 t^{(n)} = \sum_{m=0}^{n} \frac{n!}{m!(n-m)!} \{ t_{(2;1)}^{(n-m)}(1)\chi(2) + \chi_{(2;1)}(1)t^{(m)}(2) \}$$

$$t_{(2;2)}^{(n)} = \partial_2^2 t^{(n)} = \sum_{m=0}^{n} \frac{n!}{m!(n-m)!} \{ t^{(n-m)}(1)\chi_{(2;2)}(2) + \chi(1)t_{(2;2)}^{(m)}(2) \}$$

We need the quantities $T^{(n)}$ for the Schrödinger-Riccati equation and the quantities

$$T_{(1;1)}^{(n)} = \partial_1 T^{(n)} \qquad T_{(1;2)}^{(n)} = \partial_2 T^{(n)} \qquad T_{(2;1)}^{(n)} = \partial_1^2 T^{(n)} \qquad T_{(2;2)}^{(n)} = \partial_2^2 T^{(n)}$$

for the evaluation of $\partial_1 \varphi$, $\partial_2 \varphi$, $\partial_1^2 \varphi$, etc. We first present in detail the procedure for the evaluation of the quantities $T^{(n)}$. We introduce the notation

$$T_0^{(n)} \equiv T^{(n)} \qquad\qquad T_m^{(n)} = \frac{\partial^m T_0^{(n)}}{\partial \zeta^m}$$

and note that

$$T_1^{(n-1)} = \frac{\partial T_0^{(n-1)}}{\partial \zeta} = \frac{\partial T_0^{(n-1)}}{\partial \phi^{(0)}} \frac{\partial \phi^{(0)}}{\partial \zeta} = T_0^{(n)} \phi^{(1)} \equiv T^{(n)} \phi^{(1)} \tag{11.5}$$

Starting from

$$t^{(0)} = T^{(0)}$$

and taking into account Eq. (11.5) we obtain

$$t^{(1)} = \frac{\partial T^{(0)}}{\partial \zeta} = T_1^{(0)} = T_0^{(1)} \phi^{(1)} \tag{11.6}$$

This equation, from which one can obtain $T^{(1)}$, constitutes also the starting point for the recursive evaluation of the successive $T^{(n)}$, as follows.
  Successive differentiation of Eq. (11.6) yields

$$t^{(2)} = \frac{\partial t^{(1)}}{\partial \zeta} = \frac{\partial}{\partial \zeta} \{T_0^{(1)} \phi^{(1)}\} = \phi^{(2)} T_0^{(1)} + \phi^{(1)} T_1^{(1)} \tag{11.7a}$$

$$t^{(3)} = \frac{\partial^2 t^{(1)}}{\partial \zeta^2} = \frac{\partial^2}{\partial \zeta^2} \{T_0^{(1)} \phi^{(1)}\} = \phi^{(3)} T_0^{(1)} + 2\phi^{(2)} T_1^{(1)} + \phi^{(1)} T_2^{(1)} \tag{11.7b}$$

$$t^{(4)} = \frac{\partial^3 t^{(1)}}{\partial \zeta^3} = \frac{\partial^3}{\partial \zeta^3} \{T_0^{(1)} \phi^{(1)}\} = \phi^{(4)} T_0^{(1)} + 3\phi^{(3)} T_1^{(1)} + 3\phi^{(2)} T_2^{(1)} + \phi^{(1)} T_3^{(1)} \tag{11.7c}$$

$$t^{(5)} = \frac{\partial^4 t^{(1)}}{\partial \zeta^4} = \frac{\partial^4}{\partial \zeta^4} \{T_0^{(1)} \phi^{(1)}\} = \phi^{(5)} T_0^{(1)} + 4\phi^{(4)} T_1^{(1)} + 6\phi^{(3)} T_2^{(1)} + 4\phi^{(2)} T_3^{(1)}$$
$$+ \phi^{(1)} T_4^{(1)} \tag{11.7d}$$

. . .

The quantities $T_m^{(1)}$ appearing in these equations may be obtained by successive differentiation with respect to ζ of Eq. (11.5), with $n = 2$. We obtain

$$T_1^{(1)} = T_0^{(2)}\phi^{(1)} \tag{11.8a}$$

$$T_2^{(1)} = \frac{\partial T_1^{(1)}}{\partial\zeta} = \frac{\partial}{\partial\zeta}\{T_0^{(2)}\phi^{(1)}\} = \phi^{(2)}T_0^{(2)} + \phi^{(1)}T_1^{(2)} \tag{11.8b}$$

$$T_3^{(1)} = \frac{\partial^2 T_1^{(1)}}{\partial\zeta^2} = \frac{\partial^2}{\partial\zeta^2}\{T_0^{(2)}\phi^{(1)}\} = \phi^{(3)}T_0^{(2)} + 2\phi^{(2)}T_1^{(2)} + \phi^{(1)}T_2^{(2)} \tag{11.8c}$$

$$T_4^{(1)} = \frac{\partial^3 T_1^{(1)}}{\partial\zeta^3} = \frac{\partial^3}{\partial\zeta^3}\{T_0^{(2)}\phi^{(1)}\} = \phi^{(4)}T_0^{(2)} + 3\phi^{(3)}T_1^{(2)} + 3\phi^{(2)}T_2^{(2)} + \phi^{(1)}T_3^{(2)} \tag{11.8d}$$

. . .

The quantities $T_m^{(2)}$ appearing in these equations may also be obtained by successive differentiation with respect to ζ of Eq. (11.5), with $n = 3$. We obtain

$$T_1^{(2)} = T_0^{(3)}\phi^{(1)} \tag{11.9a}$$

$$T_2^{(2)} = \frac{\partial T_1^{(2)}}{\partial\zeta} = \frac{\partial}{\partial\zeta}\{T_0^{(3)}\phi^{(1)}\} = \phi^{(2)}T_0^{(3)} + \phi^{(1)}T_1^{(3)} \tag{11.9b}$$

$$T_3^{(2)} = \frac{\partial^2 T_1^{(2)}}{\partial\zeta^2} = \frac{\partial^2}{\partial\zeta^2}\{T_0^{(3)}\phi^{(1)}\} = \phi^{(3)}T_0^{(3)} + 2\phi^{(2)}T_1^{(3)} + \phi^{(1)}T_2^{(3)} \tag{11.9c}$$

. . .

Appropriate substitutions will then yield

$$t^{(2)} = \phi^{(2)}T_0^{(1)} + \phi^{(1)}(T_0^{(2)}\phi^{(1)}) = \phi^{(2)}T^{(1)} + (\phi^{(1)})^2 T^{(2)} \tag{11.10a}$$

$$t^{(3)} = \phi^{(3)}T_0^{(1)} + 2\phi^{(2)}(T_0^{(2)}\phi^{(1)}) + \phi^{(1)}(\phi^{(2)}T_0^{(2)} + \phi^{(1)}T_1^{(2)})$$

$$= \phi^{(3)}T_0^{(1)} + 2\phi^{(1)}\phi^{(2)}T_0^{(2)} + \phi^{(1)}\phi^{(2)}T_0^{(2)} + (\phi^{(1)})^2 T_1^{(2)}$$

$$= \phi^{(3)}T_0^{(1)} + 3\phi^{(1)}\phi^{(2)}T_0^{(2)} + (\phi^{(1)})^2(T_0^{(3)}\phi^{(1)})$$

$$= \phi^{(3)}T^{(1)} + 3\phi^{(1)}\phi^{(2)}T^{(2)} + (\phi^{(1)})^3 T^{(3)}$$

(11.10b)

$$t^{(4)} = \phi^{(4)}T_0^{(1)} + 3\phi^{(3)}(T_0^{(2)}\phi^{(1)}) + 3\phi^{(2)}(\phi^{(2)}T_0^{(2)} + \phi^{(1)}T_1^{(2)})$$

$$+ \phi^{(1)}(\phi^{(3)}T_0^{(2)} + 2\phi^{(2)}T_1^{(2)} + \phi^{(1)}T_2^{(2)})$$

$$= \phi^{(4)}T_0^{(1)} + 3\phi^{(1)}\phi^{(3)}T_0^{(2)} + 3(\phi^{(2)})^2 T_0^{(2)} + 3\phi^{(1)}\phi^{(2)}T_1^{(2)}$$

$$+ \phi^{(1)}\phi^{(3)}T_0^{(2)} + 2\phi^{(1)}\phi^{(2)}T_1^{(2)} + (\phi^{(1)})^2 T_2^{(2)}$$

$$= \phi^{(4)}T_0^{(1)} + 4\phi^{(1)}\phi^{(3)}T_0^{(2)} + 3(\phi^{(2)})^2 T_0^{(2)} + 5\phi^{(1)}\phi^{(2)}(T_0^{(3)}\phi^{(1)})$$

$$+ (\phi^{(1)})^2(\phi^{(2)}T_0^{(3)} + \phi^{(1)}T_1^{(3)})$$

$$= \phi^{(4)}T_0^{(1)} + 4\phi^{(1)}\phi^{(3)}T_0^{(2)} + 3(\phi^{(2)})^2 T_0^{(2)} + 6(\phi^{(1)})^2\phi^{(2)}T_0^{(3)}$$

$$+ (\phi^{(1)})^3 T_1^{(3)}$$

$$= \phi^{(4)}T_0^{(1)} + 4\phi^{(1)}\phi^{(3)}T_0^{(2)} + 3(\phi^{(2)})^2 T_0^{(2)} + 6(\phi^{(1)})^2\phi^{(2)}T_0^{(3)}$$

$$+ (\phi^{(1)})^3(T_0^{(4)}\phi^{(1)})$$

(11.10c)

$$= \phi^{(4)}T^{(1)} + 4\phi^{(1)}\phi^{(3)}T^{(2)} + 3(\phi^{(2)})^2 T^{(2)} + 6(\phi^{(1)})^2\phi^{(2)}T^{(3)}$$

$$+ (\phi^{(1)})^4 T^{(4)}$$

$$\cdots$$

and one can solve these equations for $T^{(2)}, T^{(3)}, T^{(4)}, \ldots$

The same procedure, using the quantities $t^{(n)}_{(1;1)}$, $t^{(n)}_{(1;2)}$, ..., will yield the quantities $T^{(n)}_{(1;1)}$, $T^{(n)}_{(1;2)}$, ...

This formulation, although formally efficient, may not be as precise as it could be, particularly at those *PECS* at which the quantities $t^{(n)}$ may be large and the quantities $\phi^{(n)}$ (especially $\phi^{(1)}$) may be small. In such cases it may be more appropriate to proceed as follows. Once the above expressions, Eqs. (11.10), have been obtained, they should be rewritten, by successive substitutions, in terms of the quantities $t^{(n)}$ and $\phi^{(n)}$, collecting together the terms with the same denominators, $(\phi^{(1)})^m$. In addition, for the reason given below, it will be convenient to extract, as common denominator, the highest power $(\phi^{(1)})^n$ in the expression. Proceeding in this manner we would write

$$T^{(n)} = \frac{1}{(\phi^{(1)})^n} C_n$$

with

$$C_0 = t^{(0)} \tag{11.11a}$$

$$C_1 = t^{(1)} \tag{11.11b}$$

$$C_2 = t^{(2)} - \frac{1}{\phi^{(1)}} \phi^{(2)} t^{(1)} \tag{11.11c}$$

$$C_3 = t^{(3)} - \frac{1}{\phi^{(1)}} [\phi^{(3)} t^{(1)} + 3\phi^{(2)} t^{(2)}] + \frac{3}{(\phi^{(1)})^2} (\phi^{(2)})^2 t^{(1)} \tag{11.11d}$$

$$C_4 = t^{(4)} - \frac{1}{\phi^{(1)}} [\phi^{(4)} t^{(1)} + 4\phi^{(3)} t^{(2)} + 6\phi^{(2)} t^{(3)}]$$
$$+ \frac{1}{(\phi^{(1)})^2} [10\phi^{(2)}\phi^{(3)} t^{(1)} + 15(\phi^{(2)})^2 t^{(2)}] - \frac{15}{(\phi^{(1)})^3} (\phi^{(2)})^3 t^{(1)} \tag{11.11e}$$

. . .

Then, taking into account that the quantities $T^{(n)}$ are the coefficients associated with the powers φ^n, we can rewrite the Schrödinger-Riccati equation as

$$(\mathcal{V} - E)\phi + (\mathcal{V} - E)\phi^{(1)}\omega + \sum_{n=0} \frac{1}{n!} C_n \omega^n = 0$$

with $\omega = \varphi/\phi^{(1)}$, and solve for ω, from which one will then obtain φ.

11.3.3 Evaluation of the Energy Contributions

We will first write the derivatives with respect to r_1 and r_2. From Eq. (11.4) we obtain, for differentiation with respect to r_1,

$$\phi\partial_1\mathcal{V} + (\mathcal{V}-E)\phi_{(1;1)} + \phi\partial_1\mathcal{V} + (\mathcal{V}-E)\partial_1\phi + \sum_{n=0} \frac{1}{n!} T^{(n)}_{(1;1)}\phi^n$$

$$+ \partial_1\phi \sum_{n=1} \frac{1}{(n-1)!} T^{(n)}\phi^{n-1}$$

$$= \Psi\partial_1\mathcal{V} + (\mathcal{V}-E)\phi_{(1;1)} + \sum_{n=0} \frac{1}{n!} T^{(n)}_{(1;1)}\phi^n$$

$$+ \partial_1\phi\{(\mathcal{V}-E) + \sum_{n=1} \frac{1}{(n-1)!} T^{(n)}\phi^{n-1}\} = 0$$

$$\phi\partial_1^2\mathcal{V} + 2\phi_{(1;1)}\partial_1\mathcal{V} + (\mathcal{V}-E)\phi_{(2;1)} + \phi\partial_1^2\mathcal{V} + 2\partial_1\phi\partial_1\mathcal{V} + (\mathcal{V}-E)\partial_1^2\phi$$

$$+ \sum_{n=0} \frac{1}{n!} T^{(n)}_{(2;1)}\phi^n + \partial_1\phi \sum_{n=1} \frac{1}{(n-1)!} T^{(n)}_{(1;1)}\phi^{n-1}$$

$$+ \partial_1^2\phi \sum_{n=1} \frac{1}{(n-1)!} T^{(n)}\phi^{n-1} + \partial_1\phi \sum_{n=1} \frac{1}{(n-1)!} T^{(n)}_{(1;1)}\phi^{n-1}$$

$$+ (\partial_1\phi)^2 \sum_{n=2} \frac{1}{(n-2)!} T^{(n)}\phi^{n-2}$$

$$= \Psi\partial_1^2\mathcal{V} + 2\phi_{(1;1)}\partial_1\mathcal{V} + (\mathcal{V}-E)\phi_{(2;1)} + \sum_{n=0} \frac{1}{n!} T^{(n)}_{(2;1)}\phi^n$$

$$+ 2\{\partial_1\mathcal{V} + \sum_{n=1} \frac{1}{(n-1)!} T^{(n)}_{(1;1)}\phi^{n-1}\}\partial_1\phi + (\partial_1\phi)^2 \sum_{n=2} \frac{1}{(n-2)!} T^{(n)}\phi^{n-2}$$

$$+ \{(\mathcal{V}-E) + \sum_{n=1} \frac{1}{(n-1)!} T^{(n)}\phi^{n-1}\}\partial_1^2\phi = 0$$

with equivalent expressions for differentiation with respect to r_2.

In a similar fashion, but taking into account that the quantities $T^{(n)}$ are independent of R, for differentiation with respect to R we obtain

$$\phi\partial_R\mathcal{V} + \phi\partial_R\mathcal{V} + (\mathcal{V}-E)\partial_R\phi + \partial_R\phi \sum_{n=1} \frac{1}{(n-1)!} T^{(n)}\phi^{n-1}$$

$$= \Psi\partial_R\mathcal{V} + \{(\mathcal{V}-E) + \sum_{n=1} \frac{1}{(n-1)!} T^{(n)}\phi^{n-1}\}\partial_R\phi = 0$$

$$\phi\partial_R^2 \nu + \phi\partial_R^2 \nu + 2\partial_R\nu\partial_R\varphi + (\nu - E)\partial_R^2\varphi + \partial_R^2\varphi \sum_{n=1} \frac{1}{(n-1)!} T^{(n)}\varphi^{n-1}$$

$$+ (\partial_R\varphi)^2 \sum_{n=2} \frac{1}{(n-2)!} T^{(n)}\varphi^{n-2}$$

$$= \Psi\partial_R^2\nu + 2\partial_R\nu\partial_R\varphi + (\partial_R\varphi)^2 \sum_{n=2} \frac{1}{(n-2)!} T^{(n)}\varphi^{n-2}$$

$$+ \{(\nu - E) + \sum_{n=1} \frac{1}{(n-1)!} T^{(n)}\varphi^{n-1}\}\partial_R^2\varphi$$

and finally, for differentiation with respect to both r_1 and R,

$$\phi_{(1;1)}\partial_R\nu + \partial_1\phi\partial_R\nu + \partial_1\nu\partial_R\varphi + (\nu - E)\partial_{1R}\varphi + \partial_{1R}\varphi \sum_{n=1} \frac{1}{(n-1)!} T^{(n)}\varphi^{n-1}$$

$$+ \partial_R\varphi \sum_{n=1} \frac{1}{(n-1)!} T_{(1;1)}^{(n)}\varphi^{n-1} + \partial_1\phi\partial_R\varphi \sum_{n=2} \frac{1}{(n-2)!} T^{(n)}\varphi^{n-2}$$

$$= \phi_{(1;1)}\partial_R\nu + \partial_R\nu\partial_1\varphi + \{\partial_1\nu + \sum_{n=1} \frac{1}{(n-1)!} T_{(1;1)}^{(n)}\varphi^{n-1}\}\partial_R\varphi$$

$$+ \partial_1\phi\partial_R\varphi \sum_{n=2} \frac{1}{(n-2)!} T^{(n)}\varphi^{n-2}$$

$$+ \{(\nu - E) + \sum_{n=1} \frac{1}{(n-1)!} T^{(n)}\varphi^{n-1}\}\partial_{1R}\varphi = 0$$

with an equivalent expression for differentiation with respect to r_2 and R.

The numerical values of the various derivatives of φ, evaluated with the above equations, could be substituted into the expression

$$-\frac{1}{2}(\partial_1^2\varphi + \frac{2}{r_1}\,\partial_1\varphi) - \frac{1}{2}(\partial_2^2\varphi + \frac{2}{r_2}\,\partial_2\varphi) - (\partial_R^2\varphi + \frac{2}{R}\,\partial_R\varphi)$$

$$-\frac{1}{2}\cos\,\theta_1\partial_{1R} - \frac{1}{2}\cos\,\theta_2\partial_{2R} \tag{11.12}$$

in order to obtain the energy contribution from φ. Precision considerations suggest, however, that it is more appropriate to proceed differently. The expressions that one obtains for the derivatives of φ from the above equations consist of one or more fraction(s), whose denominator(s) are powers of the quantity

$$(\nu - E) + \sum_{n=1} \frac{1}{(n-1)!} T^{(n)}\varphi^{n-1}$$

A higher precision will be achieved if one substitutes the expressions of the derivatives of φ into the above expression, Eq. (11.12), collects together the terms with the same denominator, and then proceeds with the numerical evaluation.

12 References and Bibliography

12.1 References

F. Abdullaev. *Theory of Solitons in Inhomogeneous Media*. John Wiley & Sons, Chichester (1994).

F. Abdullaev, S. Darmanyan, and P. Khabibullaev. *Optical Solitons*. Springer-Verlag, Berlin (1993).

P.B. Abraham and H.E. Moses. Phys. Rev. *A22*, 1333 (1980).

M. Abramowitz and I.A. Stegun. *Handbook of Mathematical Functions*. Dover Publications, Inc., New York (1970).

L. Adamowicz. J. Chem. Phys. *89*, 313 (1988).

L. Adamowicz and E.A. McCullough. J. Chem. Phys. *75*, 2475 (1981).

L. Adamowicz and R.J. Bartlett. J. Chem. Phys. *84*, 4988 (1986).

L. Adamowicz and R.J. Bartlett. Phys. Rev. *A37*, 1 (1988).

Y. Aharonov and C.K. Au. Phys. Rev. Lett. *42*, 1582 (1979).

T. Ahlenius and P. Lindner. J. Phys. *B8*, 778 (1975).

A.L. Akhiezer and V.B. Berestetskii. *Quantum Electrodynamics*. Interscience Publishers, New York (1965).

S.A. Alexander and H.J. Monkhorst. Int. J. Quantum Chem. *32*, 361 (1987).

S.A. Alexander, R.L. Coldwell, and H.J. Monkhorst. J. Comput. Phys. *76*, 263 (1988).

Y. Alhassid and R.D. Levine. Phys. Rev. *A18*, 89 (1978).

J. Almlöf and O. Groppen. *Relativistic Effects in Chemistry*, in *Reviews in Computational Chemistry*, edited by K.B. Lipkowitz and D.B. Boyd. Wiley-VCH, New York (vol. 8; 1996).

J.A. Alonso and N.H. March. J. Chem. Phys. *78*, 1382 (1983).

M.H. Anderson, J.R. Enscher, M.R. Mathews, C.E. Wieman, and E.A. Cornell. Science *269*, 198 (1955).

V. Aquilanti and J. Avery. Chem. Phys. Lett. *267*, 1 (1997).

F. Arias de Saavedra, E. Buendia, F.J. Galvez, and I. Porras. J. Phys. *B29*, 3803 (1996).

F.V. Atkinson. Proc. Glasgow Math. Assoc. *3*, 105 (1957).

F.V. Atkinson and C.T. Fulton. Proc. Roy. Soc. Edinburgh *99*, 51 (1984).

J. Avery. Chem. Phys. Lett. *138*, 520 (1987).

J. Avery. *Hyperspherical Harmonics. Applications in Quantum Theory*. Kluwer, Dordrecht (1989).

J. Avery and Z.Y. Wen. Int. J. Quantum Chem. *22*, 717 (1982).

J. Avery and P.S. Larsen. Int. J. Quantum Chem. *22S*, 437 (1988).

J. Avery and D.R. Herschbach. Int. J. Quantum Chem. *41*, 673 (1992).

J. Avery and F. Antonsen. Int. J. Quantum Chem. *42*, 87 (1992).

J. Avery and F. Antonsen. J. Math. Chem. *19*, 289 (1996).

J. Avery and T.B. Hansen. Int. J. Quantum Chem. *60*, 201 (1996).

J. Avery, P.S. Larsen, and S. Hengyi. Int. J. Quantum Chem. *29*, 129 (1986).

J. Avery, T.B. Hansen, M. Wang, and F. Antonsen. Int. J. Quantum Chem. *57*, 401 (1996).

J.E. Avron. Ann. Phys. *131*, 73 (1981).

I. Babuska, J. Chandra, and J.E. Flaherty. *Adaptive Computational Methods for Partial Differential Equations*. SIAM, Philadelphia (1983).

L.E. Ballentine. Am. J. Phys. *55*, 785 (1987).

I.V. Barashenkov and V.G. Makhankov. Phys. Lett. *A128*, 52 (1988).

W.A. Barker and F.N. Glover. Phys. Rev. *99*, 317 (1955).

J.H. Bartlett. Phys. Rev. *51*, 661 (1937).

J.H. Bartlett. Phys. Rev. *98*, 1067 (1955).

D.R. Bates, R.K. Lidsham, and A.L. Stewart. Phil. Trans. Roy. Soc. *246*, 15 (1953).

G. Bayn and C. Pethick. Phys. Rev. Lett. *76*, 6 (1996).

A.D. Becke. J. Chem. Phys. *76*, 6037 (1982).

A.D. Becke. J. Chem. Phys. *78*, 4787 (1983).

A.D. Becke. Int. J. Quantum Chem. *27*, 585 (1985).

A.D. Becke. Phys. Rev. *A33*, 2786 (1986).

A.D. Becke. J. Chem. Phys. *88*, 2547 (1988).

A.D. Becke. J. Chem. Phys. *96*, 2155 (1992a).

A.D. Becke. J. Chem. Phys. *97*, 2193 (1992b).

A.D. Becke and R.M. Dickson. J. Chem. Phys. *89*, 2993 (1988).

A.D. Becke and R.M. Dickson. J. Chem. Phys. *92*, 3610 (1990).

R. Benesch. Can. J. Phys. *54*, 2155 (1975).

R. Benesch and V.H. Smith. In *Wave Mechanics The First Fifty Years*, edited by W.C. Price, S.S. Chissick, and T. Ravensdale. Butterworths, London (1973).

J. Benjamin. J. Chem. Phys. *85*, 5611 (1986).

T.B. Benjamin. J. Fluid Mech. *29*, 559 (1967).

M.L. Benston. Bull. Am. Phys. Soc. *10*, 102 (1965).

M.J. Berger and J. Oliger. J. Comput. Phys. *53*, 484 (1987).

M. Berrondo and J. Recamier. Int. J. Quantum Chem. *62*, 239 (1997).

R.S. Berry. Rev. Mod. Phys. *32*, 447 (1960).

G. Berthier, M. Defranceschi, J. Navaza, M. Suard, and G. Tsoucaris. J. Mol. Struct. (Theochem) *120*, 343 (1985).

G. Berthier, M. Defranceschi, and J. Delhalle. In *Numerical Determination of the Electronic Structure of Atoms, Diatomic and Polyatomic Molecules*, edited by M. Defranceschi and J. Delhalle. Kluwer Academic Publishers, Dordrecht (1989).

G. Berthier, M. Defranceschi, and J. Delhalle. *Numerical Hartree-Fock Calculation in Momentum Space for Molecules and Polymers*, in *Self-Consistent Field Theory and Applications*, edited by R. Carbo and M. Klobukowski. Elsevier Science Publishers, Amsterdam (1990).

G. Berthier, M. Defranceschi, J. Navaza, and G. Tsoucaris. Int. J. Quantum Chem. *63*, 451 (1997).

V.I. Bespalov, A.G. Litvak, and V.I. Talanov. Second All-Union Symposium on Nonlinear Optics, Nauka (1968).

N. Bessis and G. Bessis. Phys. Rev. *A42*, 1096 (1990).

N. Bessis and G. Bessis. Phys. Rev. *A44*, 5503 (1991).

N. Bessis and G. Bessis. Phys. Rev. *A46*, 6824 (1992).

N. Bessis and G. Bessis. Phys. Rev. *A50*, 4506 (1994).

N. Bessis and G. Bessis. J. Chem. Phys. *103*, 3006 (1995).

N. Bessis and G. Bessis. Phys. Rev. *A53*, 1330 (1996).

D. Bessis, E.R. Vrscay, and C.R. Handy. J. Phys. *A20*, 419 (1987).

H.A. Bethe and E.E. Salpeter. *Quantum Mechanics of One- and Two-Electron Atoms.* Springer, Berlin, and Academic Press, Inc., New York (1957).

I. Bialynicki-Birula and Z. Bialynicka-Birula. *Quantum Electrodynamics.* Pergamon Press, Oxford, and PWN-Polish Scientific Publishers, Warsaw (1975).

F. Biggs, L.M. Mendelsohn, and J.B. Mann. At. Data Nucl. Data Tables *16*, 201 (1975).

W.A. Bingel. Z. Naturforsch. *A18*, 1249 (1963).

G. Birkhoff. J. Math. Phys. *16*, 104 (1937).

J.M. Blatt, J. Comput. Phys. *1*, 382 (1967).

R.A. Bonham and H.F. Wellenstein. In *Compton Scattering: The Investigation of Electron Momentum Distributions*, edited by B.G. Williams. McGraw-Hill, New York (1977).

M. Born. *Einstein's Theory of Relativity.* Dover Publications, New York (1962).

M. Born and J.R. Oppenheimer. Ann. der Phys. *84*, 457 (1927).

M. Born and K. Huang. *Dynamical Theory of Crystal Lattices.* Oxford University Press, Oxford (1955).

N. Bourbaki. *Elements de Mathematique.* Hermann & Cie., Paris (1961).

B.H. Bransden and C.J. Joachain. *Physics of Atoms and Molecules.* Longman, London (1983).

G. Breit. Phys. Rev. *34*, 553 (1929).

G. Breit. Phys. Rev. *36*, 383 (1930).

G. Breit. Phys. Rev. *39*, 616 (1932).

C.E. Brion. Int. J. Quantum Chem. *29*, 1397 (1986).

I.N. Bronshtein and K.A. Semendyayev. *Handbook of Mathematics.* Verlag Harri Deutsch, Thun & Frankfurt/Main; Van Nostrand Reinhold, New York (1985).

D.L. Brown and L.G.M. Reyna. SIAM J. Sci. Stat. Comput. *6* (1985).

B. Buck, L.C. Biedenharn, and R.Y. Cusson. Nucl. Phys. *A317*, 215 (1979).

R.K. Bullough and P.J. Caudrey (editors). *Solitons.* Springer-Verlag, Berlin (1980).

R.K. Bullough and P.J. Caudrey. *The Soliton and Its History* in *Solitons*, edited by R.K. Bullough and P.J. Caudrey. Springer-Verlag, Berlin (1980).

P.R. Bunker. *Molecular Symmetry and Spectroscopy.* Academic Press, Inc., New York (1979).

E.H. Burhop and H.S. Massey. Proc. Roy. Soc. London *A153*, 661 (1935).

J.L. Calais, M. Defranceschi, J.G. Priat, and J. Delhalle. J. Phys. Condens. Matter *4*, 5675 (1992).

F. Calogero and A. Degasperis. *Spectral Transform and Solitons.* North-Holland, New York (1982).

F. Calogero and W. Eckhaus. Inverse Probl. *3*, 229 (1987a).

F. Calogero and W. Eckhaus. Inverse Probl. *3*, L27 (1987b).

J.R. Cash and A.D. Raptis. Comput. Phys. Commun. *33*, 299 (1984).

J.R. Cash, A.D. Raptis, and T.E. Simos. J. Comput. Phys. *91*, 413 (1990).

M. Casida. Int. J. Quantum Chem. *27*, 451 (1985).

A.F. Chalmers. *What is this thing called Science?* University of Queensland Press, St. Lucia, Queensland, Australia (1976; second edition 1982; reprinted 1991).

D.C. Champeney. *Fourier Transforms and Their Physical Applications.* Academic Press, London (1973).

S.Y. Chang, E.R. Davidson, and G. Vincov. J. Chem. Phys. *52*, 1740 (1970).

J.A. Chapman and D.P. Chong. Can. J. Chem. *48*, 2722 (1970).

H.H. Chen and C.S. Liu. Phys. Rev. Lett. *37*, 693 (1976).

H.H. Chen and C.S. Liu. Phys. Fluids *21*, 377 (1978).

D.P. Chong. J. Chem. Phys. *47*, 4907 (1967).

Z.V. Chraplyvy. Phys. Rev. *91*, 388 (1953a).

Z.V. Chraplyvy. Phys. Rev. *92*, 1310 (1953b).

P.A. Christiansen and E.A. McCullough. J. Chem. Phys. *67*, 1877 (1977).

I. Christie, D.F. Griffiths, A.R. Mitchell, and J.M. Sanz Serna. IMA J. Numer. Anal. *1*, 253 (1981).

F. Ciacci, G. Dattoli, A. Renieri, and A. Torre. Phys. Rep. *141*, 1 (1986).

E. Clementi. *Tables of Atomic Functions.* International Business Machines Corporation, New York (1965).

L. Cohen and C. Frishberg. Phys. Rev. *A13*, 927 (1976).

M.A. Collins and D.F. Pearson. J. Chem. Phys. *99*, 6756 (1993).

E.U. Condon and G.H. Shortley. *The Theory of Atomic Spectra.* Cambridge University Press, London (1964).

H. Conroy. J. Chem. Phys. *41*, 1327 (1964a).

H. Conroy. J. Chem. Phys. *41*, 1331 (1964b).

H. Conroy. J. Chem. Phys. *41*, 1336 (1964c).

H. Conroy. J. Chem. Phys. *41*, 1341 (1964d).

H. Conroy and B.L. Bruner. J. Chem. Phys. *42*, 4047 (1965).

M.J. Cooper. Rep. Prog. Phys. *48*, 415 (1985).

F.A. Cotton. *Chemical Applications of Group Theory* (2nd edition). Wiley, New York (1970).

C.A. Coulson. Proc. Cambridge Phil. Soc. *37*, 55 (1941a).

C.A. Coulson. Proc. Cambridge Phil. Soc. *37*, 74 (1941b).

C.A. Coulson and W.E. Duncanson. Proc. Cambridge Phil. Soc. *37*, 67 (1941).

C.A. Coulson and A.C. Hurley. J. Chem. Phys. *37*, 448 (1962).

R. Courant and D. Hilbert. *Methoden der mathematischen Physik.* Julius Springer, Berlin (1931).

R. Courths and S. Hüfner. Phys. Rep. *112*, 53 (1984).

S. Cowan, R.H. Enns, S.S. Rangnekar, and S.S. Sanghera. Can. J. Phys. *64*, 311 (1986).

B. d'Espagnat. *Conceptual Foundations of Quantum Mechanics.* Benjamin, Menlo Park (second edition; 1976).

C.G. Darwin. Proc. Roy. Soc. London *A118*, 654 (1928).

G. Datoli, M. Richetta, and A. Torre. Phys. Rev. *A37*, 2007 (1988).

E.R. Davidson. *Reduced Density Matrices in Quantum Chemistry.* Academic Press, Inc., New York (1976).

D.W. Davies. Theoret. Chim. Acta *9*, 330 (1968).

L. Davis. Phys. Rev. *56*, 186 (1939).

S.F. Davis and J.E. Flaherty. SIAM J. Sci. Stat. Comput. *3*, 6 (1982).

K.B. Davis, M.O. Mewes, M.R. Andrews, N.J. van Druten, D.S. Durfee, D.M. Kurn, and W. Ketterle. Phys. Rev. Lett. *75*, 3969 (1995).

A.S. Davydov. *Quantum Mechanics.* Pergamon Press, Oxford (1965).

A.S. Davydov. *Solitons in Molecular Systems.* D. Reidel Publishing Company, Dordrecht (1985).

A.S. Davydov. *Solitons in Molecular Systems* (second edition). Kluwer Academic Publishers, Dordrecht (1991).

M.A. de Moura. Phys. Rev. *A37*, 4998 (1988).

M.A. de Moura. J. Phys. *A27*, 7157 (1994).

A. de Souza Doutra. Phys. Lett. *A131*, 319 (1988).

L. de Windt, M. Defranceschi, and J. Delhalle. Theor. Chim. Acta *86*, 487 (1993a).

L. de Windt, M. Defranceschi, and J. Delhalle. Int. J. Quantum Chem. *45*, 609 (1993b).

W.J. Deal. Int. J. Quantum Chem. *35*, 513 (1989).

B.M. Deb. *Miscellaneous Application of the Hellmann-Feynman Theorem* in *The Force Concept in Chemistry*, edited by B.M. Deb. Van Nostrand Reinhold Company, New York (1981).

M. Defranceschi and J. Delhalle. Compt. Rend. *301* (II), 1405 (1985).

M. Defranceschi and J. Delhalle. Phys. Rev. *B34*, 5862 (1986).

M. Defranceschi and J. Delhalle (editors). *Numerical Determination of the Electronic Structure of Atoms, Diatomic and Polyatomic Molecules.* NATO ASI Series C271. Kluwer, Dordrecht (1989).

M. Defranceschi and G. Berthier. J. Phys. (France) *51*, 2791 (1990).

M. Defranceschi and J. Delhalle. Eur. J. Phys. *11*, 172 (1990).

M. Defranceschi, M. Suard, and G. Berthier. Compt. Rend. *299* (II), 9 (1984a).

M. Defranceschi, M. Suard, and G. Berthier. Int. J. Quantum Chem. *25*, 683 (1984b).

M. Defranceschi, L. de Windt, J.G. Fripiat, and J. Delhalle. J. Mol. Struct. (Theochem) *258*, 179 (1992).

M. Delfour, M. Fortin, and G. Payre. J. Comput. Phys. *44*, 277 (1981).

J. Delhalle and M. Defranceschi. Int. J. Quantum Chem. *21S*, 4251 (1987).

J.B. Delos and S.M. Blinder. J. Chem. Phys. *47*, 2784 (1967).

J.P. Desclaux. Comput. Phys. Commun. *9*, 31 (1975).

B.S. DeWitt and R.N. Graham. Am. J. Phys. *39*, 724 (1971).

R.M. Dickson and A.D. Becke. J. Chem. Phys. *99*, 3898 (1993).

P.A. Dirac. Proc. Roy. Soc. *A117*, 610 (1928a).

P.A. Dirac. Proc. Roy. Soc. *A118*, 361 (1928b).

P.A. Dirac. *The Principles of Quantum Mechanics.* Oxford University Press, Oxford (1958).

R.K. Dodd, J.C. Eilbeck, J.D. Gibbon, and H.C. Morris. *Solitons and Nonlinear Wave Equations.* Academic Press, London (1982).

G.F. Drukarev. J. Eksp. Theor. Phys. *19*, 247 (1949).

J.W.H. DuMond. Phys. Rev. *33*, 643 (1929).

J.W.H. DuMond. Phys. Rev. *36*, 146 (1930).

J.W.M. DuMond. Rev. Mod. Phys. *5*, 1 (1933).

W.E. Duncanson and C.A. Coulson. Proc. Cambridge Phil. Soc. *37*, 406 (1941).

W.F. Duncanson and C.A. Coulson. Proc. Phys. Soc. (London) *57*, 190 (1945).

J.L. Dunham. Phys. Rev. *34*, 438 (1929).

T.H. Dunning. J. Chem. Phys. *90*, 1007 (1989).

A.K. Dutta and R.S. Wiley. J. Math. Phys. *29*, 892 (1988).

K.G. Dyall. J. Chem. Phys. *106*, 9618 (1997).

C.E. Dykstra. J. Chem. Phys. *87*, 2806 (1987).

C.E. Dykstra, S. Liu, F. Daskalakis, J.P. Lucia, and M. Takahashi. Chem. Phys. Lett. *137*, 266 (1987).

F.J. Dyson. Phys. Rev. *75*, 486 (1949).

C. Eckart. Phys. Rev. *46*, 487 (1934).

C. Eckart. Phys. Rev. *47*, 552 (1935).

A.R. Edmonds. *Angular Momentum in Quantum Mechanics*. Princeton University Press, Princeton (1960).

M. Edwards, R.J. Dodd, C.W. Clark, P.A. Ruprecht, and K. Burnett. Phys. Rev. *A53*, R1950 (1996).

S. Ehrenson and G.D. Harp. Int. J. Quantum Chem. *7*, 1099 (1973).

G. Eilenberger. *Solitons. Mathematical Methods for Physicists*. Springer-Verlag, Berlin (1981).

L.P. Eisenhart. Ann. Math. *35*, 284 (1934).

V.L. Eletsky and V.S. Popov. Phys. Lett. *B94*, 65 (1980).

W. Elsasser. Z. Phys. *81*, 332 (1933).

I.R. Epstein. J. Chem. Phys. *53*, 4425 (1970).

I.R. Epstein. Phys. Rev. *A8*, 8 (1973)

I.R. Epstein and W.N. Lipscomb. J. Chem. Phys. *53*, 4419 (1970).

I.R. Epstein and A.C. Tanner. In *Compton Scattering: The Investigation of Electron Momentum Distributions*, edited by B.G. Williams. McGraw-Hill, New York (1977).

J.H. Epstein and S.T. Epstein. Am. J. Phys. *30*, 266 (1962).

S.T. Epstein. J. Chem. Phys. *42*, 3813 (1965).

S.T. Epstein. *The Hellmann-Feynman Theorem* in *The Force Concept in Chemistry*, edited by B.M. Deb. Van Nostrand Reinhold Company, New York (1981).

A. Erdelyi. *Asymptotic Expansions*. Dover Publications, Inc., New York (1956).

A. Erdelyi, W. Magnus, F. Oberhettinger, and F. Tricomi. *Tables of Integral Transforms*. McGraw-Hill, New York (1954).

L.D. Faddeev. *A Hamiltonian Interpretation of the Inverse Scattering Method*, in *Solitons*, edited by R.K. Bullough and P.J. Caudrey. Springer-Verlag, Berlin (1980).

L.D. Faddeev and L.A. Takhtajan. *Hamiltonian Methods in the Theory of Solitons*. Springer-Verlag, Berlin (1987).

E.B. Fel'dman. Phys. Lett. *A104*, 479 (1984).

F. Fer. Bull. Classe Sci. Acad. Roy. Belg. *44*, 5 (1958a).

F. Fer. Bull. Classe Sci. Acad. Roy. Belg. *44*, 818 (1958b).

F.M. Fernandez. J. Math. Phys. *28*, 2908 (1987).

F.M. Fernandez. J. Chem. Phys. *88*, 490 (1988).

F.M. Fernandez. Phys. Rev. *A40*, 41 (1989).

F.M. Fernandez. Phys. Rev. *A41*, 7066 (1990).

F.M. Fernandez. Phys. Lett. *A166*, 173 (1992).

F.M. Fernandez. J. Chem. Phys. *103*, 6581 (1995a).

F.M. Fernandez. J. Phys. *A28*, 4043 (1995b).

F.M. Fernandez. J. Phys. *A29*, 3167 (1996).

F.M. Fernandez and E.A. Castro. *Hypervirial Theorems*. Springer-Verlag, Berlin (1987).

F.M. Fernandez and R. Guardiola. J. Phys. *A26*, 7169 (1993).

F.M. Fernandez and R. Guardiola. J. Phys. *A30*, 5825 (1997).

F.M. Fernandez, J. Echave, and E.A. Castro. Chem. Phys. *117*, 101 (1987).

F.M. Fernandez, E. Castro, and J.F. Ogilvie. Int. J. Quantum Chem. *36*, 61 (1989a).

F.M. Fernandez, G.I. Frydman, and E.A. Castro. J. Phys. *A22*, 64 (1989b).

F.M. Fernandez, Q. Ma, and R.H. Tipping. Phys. Rev. *A39*, 1605 (1989c).

F.M. Fernandez, Q. Ma, and R.H. Tipping. Phys. Rev. *A40*, 6149 (1989d).

F.M. Fernandez, Q. Ma, D.J. de Smet, and R.H. Tipping. Can. J. Phys. *67*, 931 (1989e).

F.H. Fernandez, J. Echave, and E.A. Castro. J. Math. Phys. *31*, 338 (1990).

F.M. Fernandez, R. Guardiola, and M. Znojil. Phys. Rev. *A48*, 417 (1993).

R.P. Feynman. Phys. Rev. *56*, 340 (1939).

R.P. Feynman. Phys. Rev. *84*, 108 (1951).

P. Fischer and M. Defranceschi. Appl. Comput. Harm. Anal. *1*, 232 (1994).

P. Fischer and M. Defranceschi. SIAM J. Numer. Anal. *35*, 1 (1998).

G. Fix. J. Math. Anal. Appl. *19*, 519 (1967).

G.P. Flessas. Phys. Lett. *A72*, 289 (1979).

G.P. Flessas. Phys. Lett. *A83*, 121 (1981).

G.P. Flessas. J. Phys. *A15*, L97 (1982).

G.P. Flessas and K.P. Das. Phys. Lett. *A78*, 19 (1980).

V. Fock. Z. Physik *63*, 855 (1930).

V. Fock. Z. Phys. *98*, 145 (1935).

L.L. Foldy. Pure Appl. Phys. *10*, 1 (1962).

L.L. Foldy and S.A. Wouthuysen. Phys. Rev. *78*, 29 (1950).

A.P. Fordy (editor). *Soliton Theory: A Survey of Results.* Manchester University Press, Manchester (1990).

A.P. Fordy. *Soliton Theory: A Brief Synopsis*, in *Soliton Theory: A Survey of Results*, edited by A.P. Fordy. Manchester University Press, Manchester (1990).

E.S. Fraga. *Adaptive Mesh Refinement Techniques for Nonlinear Dispersive Wave Equations.* Thesis, University of Waterloo, Waterloo (1988).

E.S. Fraga and J. Ll. Morris. J. Comput. Phys. *101*, 94 (1992).

E.S. Fraga and J. Ll. Morris. *A Piecewise Uniform Adaptive Grid Algorithm for Nonlinear Dispersive Wave Equations* in *Numerical Analysis: A.R. Mitchell 75th Birthday Volume*, edited by D.F. Griffiths and G.A. Watson. World Scientific Publishing Company, Singapore (1996).

S. Fraga. Semiempirical Analysis of Atomic Energy Levels. Technical Report TC-AEL-1-78, Department of Chemistry, University of Alberta, Edmonton, Alberta (1978).

S. Fraga. Phys. Rev. *A19*, 31 (1979a).

S. Fraga. Can. J. Phys. *56*, 836 (1979b).

S. Fraga. Can. J. Phys. *58*, 544 (1980).

S. Fraga. Can. J. Phys. *59*, 1668 (1981a).

S. Fraga. Anales Fis. *77*, 39 (1981b).

S. Fraga and G. Malli. *Many-Electron Systems: Properties and Interactions.* W.B. Saunders Company, Philadelphia (1968).

S. Fraga and E.S. Fraga. J. Mol. Struct. (Theochem) *426*, 1 (1998a).

S. Fraga and E.S. Fraga. Polish J. Chem. *72*, 1313 (1998b).

S. Fraga, J. Karwowski, and K.M.S. Saxena. *Handbook of Atomic Data.* Elsevier Science Publishers, Amsterdam (1976; second printing 1979).

S. Fraga, M. Klobukowski, J. Muszynska, E. San Fabian, K.M.S. Saxena, J.A. Sordo, and T.L. Sordo. *Research in Atomic Structure*. Springer, Berlin (1993).

S. Francetti. Nuovo Cimento *6*, 601 (1957).

K.A. Franken and C.E. Dykstra. J. Phys. Chem. *97*, 11408 (1993).

P. Franklin. *Basic Mathematical Formulas*, in *Fundamental Formulas of Physics*, edited by D.H. Menzel. Dover Publications, Inc., New York (1960).

C. Froese Fischer. *The Hartree-Fock Method for Atoms*. Wiley, New York (1977).

A.A. Frost. J. Chem. Phys. *10*, 240 (1942).

A.A. Frost. Theoret. Chim. Acta *1*, 36 (1962).

A.A. Frost, R.E. Kellogg, and E.C. Curtis. Rev. Mod. Phys. *32*, 313 (1960).

A.A. Frost, R.E. Kellogg, B.M. Gimarc, and J.D. Scargle. J. Chem. Phys. *35*, 827 (1961).

D. Frye, A. Preiskorn, G.C. Lie, and E. Clementi. *HYCOIN: Hylleraas Configuration Interaction Method Using Gaussian Functions* in *Modern Techniques in Computational Chemistry: MOTECC-90*, edited by E. Clementi, ESCOM, Leiden (1990).

E. Fues. Ann. Physik *81*, 281 (1926).

C.T. Fulton. J. Integral Equations *4*, 163 (1982).

S.R. Gadre, S.P. Gejji, and S. Chakravorty. At. Data Nucl. Data Tables *28*, 477 (1983).

A. Galindo and P. Pascual. *Quantum Mechanics*. Springer-Verlag, Berlin (vol. 1, 1990; vol. 2, 1991).

J.A.C. Gallas. J. Phys. *A21*, 3393 (1988).

J.M. Garcia de la Vega and B. Miguel. J. Comput. Chem. *12*, 1172 (1991).

J.M. Garcia de la Vega and B. Miguel. At. Data Nucl. Data Tables *54*, 1 (1993a).

J.M. Garcia de la Vega and B. Miguel. J. Math. Chem. *14*, 219 (1993b).

J.M. Garcia de la Vega and B. Miguel. J. Mol. Struct. (Theochem) *287*, 39 (1993c).

J.M. Garcia de la Vega and B. Miguel. At. Data Nucl. Data Tables *58*, 307 (1994).

J.M. Garcia de la Vega and B.Miguel. At. Data Nucl. Data Tables *60*, 321 (1995a).

J.M. Garcia de la Vega and B. Miguel. Chem. Phys. Lett. *236*, 616 (1995b).

J.M. Garcia de la Vega and B. Miguel. J. Solid State Chem. *116*, 275 (1995c).

J.M. Garcia de la Vega and B. Miguel. J. Math. Chem. *21*, 211 (1997).

J.M. Garcia de la Vega and B. Miguel. J. Mol. Struct. (Theochem)*426*, 35 (1998a).

J.M. Garcia de la Vega and B. Miguel. Comput. Phys. Commun. *109*, 34 (1998b).

C.S. Gardner, J.M. Greene, M.D. Kruskal, and R.M. Miura. Phys. Rev. Lett. *19*, 1095 (1967).

B. Gazdy and D.A. Micha. J. Chem. Phys. *82*, 4926 (1985).

R. Gelinas, S. Doss, and K. Miller. J. Comput. Phys. *40*, 209 (1981).

W. Gellert, S. Gottwald, M. Hellwich, H. Kastner, and H. Kustner (eds.). *The VNR Concise Encyclopedia of Mathematics*. Van Nostrand Reinhold, New York (1989).

R. Gilmore. *Lie Groups, Lie Algebras and Some of Their Applications*. Wiley-Interscience, New York (1974).

B.M. Gimarc and A.A. Frost. Theoret. Chim. Acta *1*, 87 (1963a).

B.M. Gimarc and A.A. Frost. J. Chem. Phys. *39*, 1698 (1963b).

V.L. Ginzburg and L. Pitaevskii. Sov. Phys. JETP *7*, 858 (1958).

J.P. Glusker, B.K. Patterson, and M. Rossi (eds.). *Patterson and Pattersons*. Oxford University Press, New York (1986).

J.A.N.F. Gomes and J.C. Paniagua. *Quantum Electrodynamics* in *Computational Chemistry. Structure, Interactions and Reactivity*, edited by S. Fraga. Elsevier Science Publishers, Amsterdam (1992).

R.A. Gonzales, J. Eisert, I. Koltracht, M. Neumann, and G. Rawitscher. J. Comput. Phys. *134*, 134 (1997).

J. Goodisman. J. Chem. Phys. *41*, 2365 (1964a).

J. Goodisman. J. Chem. Phys. *41*, 3889 (1964b).

W. Gordon. Z. Physik *48*, 11 (1928).

D. Gottlieb and S.A. Orszag. *Numerical Analysis of Spectral Methods. Theory and Applications*. SIAM, Philadelphia (1977).

I.S. Gradshteyn and I.M. Ryzhik. *Table of Integrals, Series, and Products*. Academic Press, New York (1980).

I.P. Grant. Adv. Phys. *19*, 747 (1970).

H. Grelland. Int. J. Quantum Chem. *19*, 847 (1981).

H.H. Grelland and J. Almlöf. Int. J. Quantum Chem. *21*, 885 (1982).

D. Griffiths, A.R. Mitchell, and J. Ll. Morris. Comp. Meth. Appl. Mech. Eng. *45*, 117 (1984).

E.P. Gross. J. Math. Phys. *4*, 195 (1963).

H. Grotch and R.A. Hegstrom. Phys. Rev. *A4*, 59 (1971).

G.G. Hall. Chem. Phys. Lett. *52*, 26 (1977).

H.F. Hameka. *Advanced Quantum Chemistry.* Addison-Wesley Publishing Company, Inc., Reading (1965).

D.K. Harriss and A.A. Frost. J. Chem. Phys. *40*, 204 (1964).

D.K. Harriss and C.M. Carlson. J. Chem. Phys. *51*, 5458 (1969).

D.K. Harriss and R.K. Roubal. Theor. Chim. Acta *9*, 303 (1968).

B.J. Harris. J. Math. Analysis Appl. *137*, 462 (1989).

D.R. Hartree. *The Calculation of Atomic Structures.* Wiley, New York (1957).

A. Hasegawa. *Optical Solitons in Fibers.* Springer-Verlag, Berlin (1989).

H. Hasimoto and H. Ohno. J. Phys. Soc. Japan *33*, 805 (1972).

R.W. Hasse. Phys. Rev. *A25*, 583 (1982).

R.A. Hegstrom. Phys. Rev. *A7*, 451 (1973).

D. Heinemann, D. Kolb, and B. Fricke. Chem. Phys. Lett. *137*, 180 (1987).

D. Heinemann, B. Fricke, and D. Kolb. Chem. Phys. Lett. *185*, 125 (1988a).

D. Heinemann, B. Fricke, and D. Kolb. Phys.Rev. *A38*, 4994 (1988b).

H. Hellmann. *Einführung in die Quantenchemie.* Franz Deuticke, Leipzig (1937).

W.H. Henneker and P.E. Cade. Chem. Phys. Lett. 2, 575 (1968).

B. Herbst, J. Ll. Morris, and A.R. Mitchell. J. Comput. Phys. *60*, 282 (1985).

R. Hermann. *Prolongations, Bäcklund Transformations, and Lie Theory as Algorithms for Solving and Understanding Nonlinear Differential Equations,* in *Solitons in Action,* edited by K. Lonngren and A. Scott. Academic Press, New York (1978).

J.D. Hey. Am. J. Phys. *61*, 28 (1993).

K. Hirao. Kagaku *51*, 37 (1996).

J.O. Hirschfelder. J. Chem. Phys. *33*, 1462 (1960).

J.O. Hirschfelder and E. Wigner. Proc. Natl. Acad. Sci. USA *21*, 113 (1935).

J.O. Hirschfelder and C.A. Coulson. J. Chem. Phys. *36*, 941 (1962).

J.O. Hirschfelder and M.A. Eliason. J. Chem. Phys. *47*, 1164 (1967).

D. Hoang Binh and H. Van Regemorter. J. Phys. *B30*, 2403 (1997).

H. Hochstadt. Comm. Pure Appl. Math. 749 (1961).

P. Hohenberg and W. Kohn. Phys. Rev. B *136*, 864 (1964).

M. Holland and J. Cooper. Phys. Rev. *A53*, R1954 (1996).

W.V. Houston. *Principles of Quantum Mechanics.* Dover Publications, Inc., New York (1959).

E. Humphreys. *Introduction to Lie Algebras and Representation Theory.* Springer-Verlag, Berlin (1972).

A.C. Hurley. *The Molecular Orbital Interpretation of Bond-Length Changes Following Excitation and Ionization of Diatomic Molecules*, in *Molecular Orbitals in Chemistry, Physics, and Biology*, edited by P.-O. Löwdin and B. Pullman. Academic Press, New York (1964).

K. Husimi. Prog. Theoret. Phys. *9*, 381 (1953).

E.A. Hylleraas. Z. Physik *54*, 347 (1929).

E.A. Hylleraas. Z. Phys. *74*, 216 (1932).

T. Imbo and U. Sukhatme. Am. J. Phys. *52*, 140 (1984).

E.L. Ince. *Ordinary Differential Equations.* Dover, New York (1956).

L. Infeld and T.D. Hull. Rev. Mod. Phys. *23*, 21 (1951).

J.M. Iñiguez Almech. *Mecánica Cuántica.* Memorias de la Academia de Ciencias de Zaragoza, Serie 2, Memoria 2, Zaragoza (1949).

Y. Ishikawa, W. Rodriguez, and S.A. Alexander. Int. J. Quantum Chem. *215*, 417 (1987).

Y. Ishikawa, W. Rodriguez, and S. Torres. Chem. Phys. Lett. *143*, 289 (1988).

L.G. Ixaru. Comput. Phys. Commun. *20*, 97 (1980).

L.G. Ixaru. *Numerical Methods for Differential Equations and Applications.* Reidel, Dordrecht (1984).

L.G. Ixaru and M. Rizea. Comput. Phys. Commun. *19*, 23 (1980).

N. Jacobson. *Lie Algebras.* Interscience Publishers, Inc., New York (1962).

E. Jahnke and F. Emde. *Tables of Functions with Formulae and Curves.* Dover Publications, New York (1945).

M. Jammer. *The Philosophy of Quantum Mechanics.* Wiley, New York (1974).

W.J. Janis, P. Kaijser, V.H. Smith, and M. Whangbo. Mol. Phys. *35*, 1237 (1978).

W.J. Janis, P. Kaijser, J.R. Sabin, and V.H. Smith. Mol. Phys. *37*, 463 (1979).

F. Javor, G.F. Thomas, and S.M. Rothstein. Int. J. Quantum Chem. *11*, 59 (1977).

R.S. Johnson. Proc. Roy. Soc. (London) *A375*, 131 (1977).

B.R. Judd. *Angular Momentum Theory for Diatomic Molecules.* Academic Press, New York (1975).

T. Kahan. *Theorie des Groupes en Physique Classique et Quantique.* Dunod, Paris (1960).

P. Kaijser and P. Lindner. Phil. Mag. *31*, 871 (1975).

P. Kaijser and V.H. Smith. Mol. Phys. *31*, 1557 (1976a).

P. Kaijser and V.H. Smith. *On Inversion Symmetry in Momentum Space*, in *Quantum Science: Methods and Structure*, edited by J.L. Calais, O. Goscinski, J. Linderberg, and Y. Öhrn. Plenum Press, New York (1976b).

P. Kaijser and V.H. Smith. Adv. Quantum Chem. *10*, 371 (1977).

P. Kaijser, P. Lindner, A. Andersen, and E.W. Thulstrup. Chem. Phys. Lett. *23*, 409 (1973).

P. Kaijser, V.H. Smith, and A.J. Thakkar. Mol. Phys. *41*, 1143 (1980).

T. Kakutani and K. Michihiro. J. Phys. Soc. Japan *52*, 4129 (1983).

I.G. Kaplan. *Symmetry of Many-Electron Systems*. Academic Press, Inc., New York (1975).

T. Kato. Trans. Amer. Math. Soc. *70*, 195 (1951).

T. Kato. Commun. Pure Appl. Math. *10*, 151 (1957).

D. Kaup and A.C. Newell. J. Math. Phys. *19*, 798 (1978).

J.D. Kelley. J. Chem. Phys. *56*, 6108 (1972).

P.L. Kelley. Phys. Rev. Lett. *15*, 1005 (1965).

E.C. Kemble. *The Fundamental Principles of Quantum Mechanics with Elementary Applications*. Dover Publications, Inc., New York (1958).

R.A. Kendall, T.H. Dunning, and R.J. Harrison. J. Chem. Phys. *96*, 6796 (1992).

E.H. Kerner. Can. J. Phys. *36*, 371 (1958).

G.E. Kilby. Proc. Phys. Soc. (London) *86*, 1037 (1965).

J. Killinbeck. Phys. Lett. *A67*, 13 (1978).

J.C. Kimball. Phys. Rev. *A7*, 1648 (1973).

J.C. Kimball. J. Phys. *A8*, 1513 (1975).

F.W. King and S.M. Rothstein. Phys. Rev. *A21*, 1378 (1980).

F.W. King and B.D. Dalke. J. Chem. Phys. *78*, 3143 (1983).

F.W. King and M.E. Poitzsch. Mol. Phys. *51*, 835 (1984).

F.W. King, M.A. LeGore, and M.K. Kelly. J. Chem. Phys. *75*, 809 (1981).

F.W. King, M.K. Kelly, and M.A. LeGore. J. Chem. Phys. *76*, 574 (1982).

F.W. King, L.G. Nemec, and M.K. Kelly. Mol. Phys. *50*, 1285 (1983).

F.W. King, M.K. Kelly, M.A. LeGore, and M.E. Poitzsch. Comput. Phys. Commun. *32*, 215 (1984).

F.W. King, K.J. Dykema, and B.D. Dalke. J. Chem. Phys. *96*, 2889 (1992*a*).

F.W. King, S.E. Kelly, and M.A. Kuehne. Mol. Phys. *75*, 243 (1992*b*).

H.F. King. Theor. Chim. Acta *94*, 345 (1996).

J.R. Klauder and E.C.G. Sudarshan. *Fundamentals of Quantum Optics.* W.A. Benjamin, Inc., New York (1968).

J. Kobus. Chem. Phys. Lett. *202*, 7 (1993).

J. Kobus. Comput. Phys. Commun. *78*, 247 (1994).

J. Kobus, D. Moncrieff, and S. Wilson. J. Phys. *B27*, 5139 (1994).

J. Kobus, D. Moncrieff, and S. Wilson. Mol. Phys. *86*, 1315 (1995).

J. Kobus, L. Laaksonen, and D. Sundholm. Comput. Phys. Commun. *98*, 346 (1996).

T. Koga. J. Chem. Phys. *82*, 2022 (1985a).

T. Koga. J. Chem. Phys. *83*, 2328 (1985b).

T. Koga and M. Morita. Theoret. Chim. Acta *59*, 423 (1981).

T. Koga and S. Matsumoto. J. Chem. Phys. *82*, 5127 (1985).

T. Koga and M. Ujiie. J. Chem. Phys. *84*, 335 (1986).

T. Koga and R. Kawa-ai. J. Chem. Phys. *84*, 5651 (1986).

T. Koga and T. Matsuhashi. J. Chem. Phys. *87*, 1677 (1987).

T. Koga and T. Matsuhashi. J. Chem. Phys. *89*, 983 (1988).

T. Koga and T. Ougihara. J. Chem. Phys. *91*, 1092 (1989).

T. Koga and H. Yamazaki. J. Chem. Phys. *91*, 3020 (1989).

T. Koga and A.J. Thakkar. J. Phys. *B29*, 2973 (1996).

T. Koga, T. Uchiyama, and M. Ujiie. J. Chem. Phys. *87*, 4025 (1987).

T. Koga, Y. Yamamoto, and T. Matsuhashi. J. Chem. Phys. *88*, 6675 (1988).

T. Koga, T. Horiguchi, and Y. Yshikawa. J. Chem. Phys. *95*, 1086 (1991).

W. Kolos and C.C.J. Roothaan. Rev. Mod. Phys. *32*, 205 (1960).

W. Kolos and L. Wolniewicz. J. Chem. Phys. *45*, 509 (1966a).

W. Kolos and L. Wolniewicz. J. Chem. Phys. *45*, 944 (1966b).

F.F. Komarov and M.M. Temkin. J. Phys. *B9*, L255 (1976).

Z. Kopal. *Numerical Analysis.* Chapman and Hall, London (1955). John Wiley & Sons, Inc., New York (2nd edition; 1961).

D.J. Korteweg and G. de Vries. Phil. Mag. Ser. 5, *39*, 422 (1895).

M. Kotani, A. Amemiya, E. Ishiguro, and T. Kimura. *Tables of Molecular Integrals.* Maruzen, Tokyo (1955).

M.D. Kruskal. *Nonlinear Wave Equations*, in *Dynamical Systems. Theory and Applications*, edited by J. Moser. Springer-Verlag, Berlin (1975).

R. Kubo. J. Chem. Phys. *20*, 770 (1952).

S.A. Kulkarni and S.R. Gadre. Z. Naturforsch. *48A*, 145 (1993).

S.A. Kulkarni, S.R. Gadre, and R.K. Pathak. Phys. Rev. *A45*, 4399 (1992).

A. Kundu. Physica *D25*, 399 (1987).

W. Kutzelnigg and W. Klopper. J. Chem. Phys. *94*, 1985 (1991).

L. Laaksonen and I.P. Grant. Chem. Phys. Lett. *109*, 485 (1984a).

L. Laaksonen and I.P. Grant. Chem. Phys. Lett. *112*, 157 (1984b).

L. Laaksonen, P. Pyykkö, and D. Sundholm. Comput. Phys. Rep. *4*, 313 (1986).

L. Laaksonen, F. Müller-Plathe, and G.H.F. Diercksen. J. Chem. Phys. *89*, 4903 (1988).

G.L. Lamb. *Elements of Soliton Theory*. Wiley, Toronto (1980).

L.D. Landau and E.M. Lifshitz. Mecanica Cuantica No-Relativista. Editorial Reoerte, S.A., Barcelona (1983).

S.R. Langhoff and R.A. Tawil. J. Chem. Phys. *63*, 2745 (1975).

E.N. Lasettre. J. Chem. Phys. *58*, 1991 (1973).

E.N. Lasettre. In *Numerical Determination of the Electronic Structure of Atoms, Diatomic and Polyatomic Molecules*, edited by M. Defranceschi and J. Delhalle. Kluwer, Dordrecht (1989).

P.G. Leach, G.P. Flessas, and V.M. Gorringa. J. Math. Phys. *30*, 406 (1989).

R.A. Leacock and M.J. Padgett. Phys. Rev. Lett. *50*, 3 (1983).

C.F. Lebeda and W.R. Thorson. Can. J. Chem. *48*, 2937 (1970).

K.T. Leung and C.E. Brion. Chem. Phys. *96*, 241 (1985).

I.N. Levine. *Quantum Chemistry*. Prentice Hall, Englewood Cliffs (1991; fourth edition).

M. Levy. Proc. Roy. Soc. (London) *204A*, 1451 (1950).

S. Liu and C.E. Dykstra. J. Phys. Chem. *90*, 3097 (1986).

S. Liu, R.G. Parr, and A. Nagy. Phys. Rev. *A52*, 2645 (1995).

C.Y. Loh. *The Grid Spatial Stability and an Adaptive Scheme for the Diffusion-Convection Equations*. Research Report CS-87-43, Faculty of Mathematics, University of Waterloo, Waterloo (1987).

H.C. Longuet-Higgins. Mol. Phys. *6*, 445 (1963).

K. Lonngren and A. Scott (editors). *Solitons in Action*. Academic Press, New York (1978).

R. Loudon. *The Quantum Theory of Light*. Clarendon Press, Oxford (second edition; 1983).

P.-O. Löwdin. Phys. Rev. *97*, 1474 (1955).

P.-O. Löwdin. Mol. Spectrosc. *5*, 46 (1959).

M.B. Luban and D.L. Pursey. Phys. Rev. *D33*, 431 (1986).

J.K.L. MacDonald. Phys. Rev. *46*, 828 (1934).

W. Magnus. Commun. Pure Appl. Math. *7*, 649 (1954).

W. Magnus, F. Oberhettinger, and R.P. Soni. *Formulas and Theorems for the Special Functions of Mathematical Physics*. Springer-Verlag, New York (third edition; 1966).

E. Magyari. Phys. Lett. *A81*, 116 (1981).

V.S. Manoranjan. *An Adaptive Scheme in One Space Dimension*. Report NA/76, Department of Mathematics, University of Dundee, Dundee (1984).

Maple V. Waterloo Maple Software. 450 Phillip St., Waterloo, Ontario, Canada N2L 5J2.

M.M. Maricq. Phys. Rev. *B25*, 6622 (1982).

M.M. Maricq. J. Chem. Phys. *85*, 5167 (1986).

J. Maruani and J. Serre (eds.). *Symmetries and Properties of Non-Rigid Molecules*. Elsevier Science Publishers, Amsterdam (1983).

R.L. Matcha and B.M. Pettit. J. Chem. Phys. *70*, 3130 (1979).

J. Mathews and R.L. Walker. *Mathematical Methods of Physics*. W.A. Benjamin, Inc., New York (1964); Benjamin/Cummings Publishing Company, Menlo Park (1970).

I.E. McCarthy and E. Weigold. Rep. Prog. Phys. *54*, 789 (1991).

A.J. McConnell. *Applications of Tensor Analysis*. Dover Publications, Inc., New York (1957).

E.A. McCullough. Chem. Phys. Lett. *24*, 55 (1974).

E.A. McCullough. J. Chem. Phys. *62*, 3991 (1975).

E.A. McCullough. J. Chem. Phys. *75*, 1579 (1981).

E.A. McCullough. Comput. Phys. Rep. *4*, 265 (1986).

R. McWeeny. Proc. Phys. Soc. *A62*, 519 (1949).

R. McWeeny and C.A. Coulson. Proc. Phys. Soc. *A62*, 509 (1949).

R. McWeeny. *Methods of Molecular Quantum Mechanics*. Academic Press, London (second edition; 1989).

L.B. Mendelsohn and V.H. Smith. *Atoms*, in *Compton Scattering*, edited by B.G. Williams. McGraw-Hill, New York (1977).

H.H. Michels. J. Chem. Phys. *44*, 3834 (1966).

B. Mielnick and M.A. Reyes. J. Phys. *A29*, 6009 (1996).

K. Miller and R. Miller. SIAM J. Numer. Anal. *18*, 1019 (1981).

W.E. Milne. Phys. Rev. *35*, 863 (1930).

A.R. Mitchell and D.F. Griffiths. *The Finite Difference Method in Partial Differential Equations.* John Wiley and Sons, London (1980).

A.R. Mitchell and J. Ll. Morris. Arab Gulf Journal of Scientific Research *1*, 461 (1983).

R.M. Miura. SIAM Review *18*, 412 (1976).

R. Miura. *An Introduction to Solitons and the Inverse Scattering Method via the Korteweg-de Vries Equation*, in *Solitons in Action*, edited by K. Lonngren and A. Scott. Academic Press, New York (1978).

R.M. Miura, C.S. Gardner, and M.D. Kruskal. J. Math. Phys. *9*, 1204 (1968).

D. Moncrieff, J. Kobus, and S. Wilson. J. Phys. *B28*, 4555 (1995).

H.J. Monkhorst and B. Jeziorski. J. Chem. Phys. *71*, 5268 (1979).

J.H. Moore, J.A. Tossel, and M.A. Coplan. Acc. Chem. Res. *15*, 192 (1982).

D.J. Morgan and P.T. Landsberg. Proc. Phys. Soc. London *86*, 261 (1965).

S.A. Morgan, R.J. Ballagh, and K. Burnett. Phys. Rev. *A55*, 4338 (1997).

M.C. Mosher. J. Comput. Phys. *57*, 157 (1985).

R.E. Moss. Am. J. Phys. *39*, 1169 (1971).

R.E. Moss. *Advanced Molecular Quantum Mechanics.* Chapman and Hall, London (1973).

N.F. Mott and I.N. Sneddon. *Wave Mechanics and Its Applications.* Dover Publications, Inc., New York (1963).

F. Müller-Plathe and L. Laaksonen. Chem. Phys. Lett. *160*, 175 (1989).

H. Nakatsuji. Phys. Rev. *A14*, 41 (1976).

A.B. Nassar. Phys. Rev. *A33*, 3502 (1986).

J. Navaza and G. Tsoucaris. Phys. Rev. *A24*, 683 (1981).

B. Nelander. J. Chem. Phys. *51*, 469 (1969).

W.I. Newman and W.R. Thorson. Can. J. Phys. *50*, 2997 (1972).

Y. Nogami and F.M. Toyama. Phys. Rev. *E49*, 4497 (1994a).

Y. Nogami and F.M. Toyama. Phys. Lett. *A184*, 245 (1994b).

J.H. Noggle. *Physical Chemistry.* Little, Brown and Company. Boston (1985).

B.K. Novosadov. Opt. Spectrosc. *41*, 490 (1976).

B.K. Novosadov and A.I. Pogonin. Theoret. Chim. Acta *60*, 495 (1982).

H. Ohno. J. Phys. Soc. Japan *39*, 1082 (1975).

P.J. Olver and D.H. Sattinger. *Solitons in Physics, Mathematics, and Nonlinear Optics.* Springer-Verlag, Berlin (1990).

J.R. Oppenheimer. Zeits. Physik *41*, 268 (1927).

J.R. Oppenheimer. Phys. Rev. *31*, 66 (1928).

R.T. Pack and W. Byers Brown. J. Chem. Phys. *45*, 556 (1966).

W.E. Palke. Theoret. Chim. Acta *71*, 401 (1987).

C.D. Papageorgiou and A.D. Raptis. Comput. Phys. Commun. *43*, 325 (1987).

C.D. Papageorgiou, A.D. Raptis, and T.E. Simos. J. Comput. Appl. Math. *29*, 61 (1990a).

C.D. Papageorgiou, A.D. Raptis, and T.E. Simos. J. Comput. Phys. *88*, 477 (1990b).

C.A. Parish and C.E. Dykstra. J. Phys. Chem. *97*, 9374 (1993).

R.G. Parr and J.E. Brown. J. Chem. Phys. *49*, 4849 (1968).

R.G. Parr and R.G. Pearson. J. Am. Chem. Soc. *105*, 7512 (1983).

R.G. Parr and W. Yang. *Density Functional Theory of Atoms and Molecules.* Oxford University Press, New York (1989).

R.G. Parr, R.A. Donnelly, M. Levy, and W.E. Palke. J. Chem. Phys. *68*, 3901 (1978).

R.G. Parr and W. Yang. Ann. Rev. Phys. Chem. *46*, 701 (1995).

R.K. Pathak, B.S. Sharma, and A.J. Thakkar. J. Chem. Phys. *85*, 958 (1986).

D. Pathria and J. Ll. Morris. Physica Scripta *39*, 673 (1989).

D. Pathria and J. Ll. Morris. J. Comput. Phys. *87*, 108 (1990).

D. Pathria and J.Ll. Morris. Appl. Numer. Math. *8*, 243 (1991).

W. Pauli. Z. Physik *43*, 601 (1927).

L. Pauling and E. Bright Wilson, Jr. *Introduction to Quantum Mechanics.* McGraw-Hill Book Company, Inc., New York (1935).

R. Pauncz. *Spin Eigenfunctions: Construction and Use.* Plenum, New York (1979).

R.G. Pearson and W.F. Palke. Int. J. Quantum. Chem. *37*, 103 (1990).

P. Pechukas and J.C. Light. J. Chem. Phys. *44*, 3897 (1966).

J.P. Perdew. Phys. Rev. *B33*, 88 (1986a).

J.P. Perdew. Phys. Rev. *B34*, 7406 (1986b)

J.P. Perdew and Y. Wang. Phys. Rev. *B45*, 13244 (1992).

C.L. Pekeris. Phys. Rev. *112*, 1649 (1958).

V. Pereyra and E.G. Sewell. Numer. Math. *23*, 261 (1975).

P. Phillipson. J. Chem. Phys. *39*, 3010 (1963).

F.L. Pilar. *Elementary Quantum Chemistry*. McGraw-Hill Book Company, New York (1968).

Ph. Pluvinage. *Eléments de Mécanique Quantique*. Masson et Cie., Editeurs, Paris (1955).

B. Podolsky and L. Pauling. Phys. Rev. *34*, 109 (1929).

S.M. Poling, E.R. Davidson, and G. Vincov. J. Chem. Phys. *54*, 3005 (1971).

P. Politzer and H. Weinstein. J. Chem. Phys. *71*, 4128 (1979).

V.H. Ponce. At. Data Nucl. Data Tables *28*, 477 (1983).

W.H. Press, S.A. Teukolsky, W.T. Vetterling, and B.P. Flannery. *Numerical Recipes in Fortran. The Art of Scientific Computing*. Cambridge University Press, Cambridge (1986; second edition 1992).

V. Privman. Phys. Lett. *A81*, 326 (1981).

H. Prüfer. Math. Ann. *95*, 409 (1926).

D.L. Pursey. Phys. Rev. *D33*, 1048 (1986a).

D.L. Pursey. Phys. Rev. *D33*, 2267 (1986b).

D.L. Pursey. Phys. Rev. *D36*, 1103 (1987).

K.I. Pushkarov, D.I. Pushkarov, and I.V. Tomov. Opt. Quantum Electron *11*, 471 (1979).

P. Pyykkö, G.H.F. Diercksen, F. Müller-Plathe, and L. Laaksonen. Chem. Phys. Lett. *141*, 535 (1987).

R. Radhakrishnan and M. Lakshmanan. Phys. Rev. *E54*, 2949 (1996).

E.D. Rainville. *Special Functions*. MacMillan, New York (1960).

A.K. Rajagopal, J.C. Kimball, and M. Banerjee. Phys. Rev. *B18*, 2339 (1978).

B.I. Ramirez. J. Phys. *B15*, 4339 (1982).

D. Rapp. *Quantum Mechanics*. Holt, Rinehart and Winston, Inc., New York (1971).

A.D. Raptis. Computing *28*, 373 (1982).

A.D. Raptis and A.C. Allison. Comput. Phys. Commun. *14*, 1 (1978).

A.D. Raptis and J.R. Cash. Comput. Phys. Commun. *36*, 113 (1985).

A.D. Raptis and J.R. Cash. Comput. Phys. Commun. *44*, 95 (1987).

V.A. Rassolov and D.M. Chipman. J. Chem. Phys. *104*, 9909 (1996).

D.C. Rawling and E.R. Davidson. J. Phys. Chem. *89*, 969 (1985).

C. Rebbi and G. Solani. *Solitons and Particles*. World Scientific, Singapore (1984).

P.E. Regier, J. Fischer, B.S. Sharma, and A.J. Thakkar. Int. J. Quantum Chem. *28*, 429 (1985).

M.A. Revilla. Int. J. Numer. Methods *23*, 2263 (1986).

K.W. Richman and E.A. McCullough. J. Phys. Chem. *92*, 2714 (1988).

H.P. Robertson. Math. Annalen *98*, 749 (1928).

D.W. Robinson. Helv. Phys. Acta *36*, 140 (1963).

M.P. Robinson. Computers Math. Applic. *23*, 39 (1997).

M.P. Robinson and G. Fairweather. Numer. Math. *68*, 355 (1994).

W. Rodriguez and Y. Ishikawa. Chem. Phys. Lett. *146*, 515 (1988a).

W. Rodriguez and Y. Ishikawa. Int. J. Quantum Chem. *22S*, 445 (1988b).

C.C.J. Roothaan and A.W. Weiss. Rev. Mod. Phys. *32*, 194 (1960).

C.C.J. Roothaan and P.S. Kelly. Phys. Rev. *131*, 1177 (1963).

T.A. Rourke and E.T. Stewart. Can. J. Phys. *45*, 2755 (1967).

T.A. Rourke and E.T. Stewart. Can. J. Phys. *46*, 1603 (1968).

P. Roy and R.K. Roychoudhury. Phys. Lett. *A122*, 275 (1987).

R.K. Roychoudhury and Y.P. Varshni. J. Phys. *A15*, 3025 (1988).

A. Rubinowicz. Phys. Rev. *73*, 1330 (1948).

P.A. Ruprecht, M. Holland, K. Burnett, and M. Edwards. Phys. Rev. *51*, 4704 (1995).

J.S. Russell. *Report on Waves*. Fourteenth Meeting of the British Association for the Advancement of Science. John Murray, London (1844).

R.D. Russell and J. Christiansen. SIAM J. Numer. Anal. *15*, 59 (1978).

D.E. Rutherford. *Substitutional Analysis*. Edinburgh University Press, Edinburgh (1948).

R.P. Sagar, A.C.T. Ku, V.H. Smith, and A.M. Simas. J. Chem. Phys. *90*, 6520 (1989).

L.D. Salem and R. Montemayor. Phys. Rev. *A43*, 1169 (1991).

E.E. Salpeter. Phys. Rev. *84*, 1226 (1951).

W.R. Salzman, J. Chem. Phys. *82*, 822 (1985).

W.R. Salzman. Chem. Phys. Lett. *124*, 531 (1986a).

W.R. Salzman. J. Chem. Phys. *85*, 4605 (1986b).

W.R. Salzman. Phys. Rev. *A36*, 5074 (1987).

J. Sanchez Mondragon and K.B. Wolf (editors). *Lie Methods in Optics*. Springer-Verlag, Berlin (1986).

J.M. Sanz Serna. J. Comput. Phys. *52*, 273 (1982).

J.M. Sanz Serna and I. Christie, J. Comput. Phys. *39*, 94 (1981).

J.M. Sanz Serna and J.G. Verwer. *Conservative and non conservative schemes for the solution of the nonlinear Schrödinger equation*. Report NM-R8405, Center for Mathematics and Computer Science, Amsterdam (1984).

J.M. Sanz Serna and I. Christie. J. Comput. Phys. *67*, 348 (1986).

D.H. Sattinger and O.L. Weaver. *Lie Groups and Algebras with Applications to Physics, Geometry and Mechanics*. Springer-Verlag, Berlin (1986).

R.P. Saxena and V.S. Varma. J. Phys. *A15*, L221 (1982).

H. Schlosser. Phys. Rev. *A15*, 1349 (1977).

A. Schmelzer and J.N. Murrell. Int. J. Quantum Chem. *28*, 288 (1985).

D.S. Schonland. *Molecular Symmetry*. Van Nostrand Reinhold, New York (1965).

E. Schrödinger. Ann. Physik *79*, 321 (1926a).

E. Schrödinger. Ann. Physik *79*, 480 (1926b).

E. Schrödinger. Ann. Physik *79*, 734 (1926c).

E. Schrödinger. Ann. Physik *79*, 748 (1926d).

E. Schrödinger. Ann. Physik *80*, 437 (1926e).

E. Schrödinger. Ann. Physik *81*, 109 (1926f).

W. Schulze and D. Kolb. Chem. Phys. Lett. *122*, 271 (1985).

M.W. Scott. J. Chem. Phys. *60*, 3875 (1974).

D.J. Searles and E.I. von Nagy-Felsobuki. Comput. Phys. Commun. *67*, 527 (1992).

B.S. Sharma, A.N. Tripathi, and A.J. Thakkar. J. Chem. Phys. *79*, 3164 (1983).

T. Shibuya and C.E. Wulfman. Proc. Roy. Soc. (London) *286A*, 3761 (1965).

H.K. Shin. J. Chem. Phys. *70*, 4285 (1983).

J.R. Silva and S. Canuto. Phys. Lett. *A88*, 282 (1982).

J.R. Silva and S. Canuto. Phys. Lett. *106*, 1 (1984a).

J.R. Silva and S. Canuto. Phys. Lett. *A101*, 326 (1984b).

H.J. Silverstone, M.L. Yin, and R.L. Somorjai. J. Chem. Phys. *47*, 4824 (1967).

A.M. Simas, A.J. Thakkar, and V.H. Smith. Int. J. Quantum Chem. *21*, 419 (1982).

A.H. Simas, W.H. Westgate, and V.H. Smith. J. Chem. Phys. *80*, 2636 (1984a).

A.M. Simas, V.H. Smith, and P. Kaijser. Int. J. Quantum Chem. *25*, 1035 (1984b).

G.F. Simmons. *Differential Equations with Applications and Historical Notes.* McGraw-Hill Book Company, New York (1972).

T.E. Simos. J. Comput. Appl. Math. *30*, 251 (1990).

T.E. Simos. IMA J. Numer. Anal. *11*, 347 (1991).

T.E. Simos. Comput. Phys. Commun. *71*, 32 (1992).

T.E. Simos and G. Tougelidis. Int. J. Quantum Chem. *59*, 477 (1996).

J.C. Slater. J. Chem. Phys. *1*, 687 (1933).

C.P. Slichter. *Principles of Magnetic Resonance.* Harper and Row, New York (1963).

V.H. Smith. Chem. Phys. Lett. *9*, 365 (1971).

I.N. Sneddon. *Fourier Transforms.* McGraw-Hill, New York (1951).

L.C. Snyder and T.A. Weber. J. Chem. Phys. *68*, 2974 (1978).

M.R. Spiegel. *Theory and Problems of Advanced Calculus.* Schaum Publishing Co., New York (1963).

R.E. Stanton and R.L. Taylor. J. Chem. Phys. *45*, 565 (1966).

E.O. Steinborn and K. Ruedenberg. *Rotation and Translation of Regular and Irregular Solid Spherical Harmonics*, in *Advances in Quantum Chemistry*, edited by P.O. Löwdin. Academic Press, New York (1973).

E. Steiner. J. Chem. Phys. *39*, 2365 (1963).

E. Steiner. J. Chem. Phys. *59*, 2427 (1973).

G. Stephenson and C.W. Kilmister. *Special Relativity for Physicists.* Dover Publications, New York (1987).

G. Strang and G.J. Fix. *An Analysis of the Finite Element Method.* Prentice-Hall, Inc., Englewood Cliffs (1973).

W.A. Strauss. *The Nonlinear Schrödinger Equation* in *Contemporary Developments in Continuum Mechanics*, edited by G.M. de la Penha and L.A. Medeiros. North Holland, New York (1978).

D. Sundholm. Chem. Phys. Lett. *149*, 251 (1988a).

D. Sundholm. Comput. Phys. Commun. *49*, 409 (1988b).

B.T. Sutcliffe. *The Born-Oppenheimer Approximation* in *Methods in Computational Molecular Physics*, edited by S. Wilson and G.H.F. Diercksen. NATO ASI, Series B, vol. 293. Plenum Press, New York (1992).

B.T. Sutcliffe. J. Chem. Soc. Faraday Trans. *89*, 2321 (1993).

B.T. Sutcliffe. *The Decoupling of Nuclear from Electronic Motions in Molecules* in *Conceptual Trends in Quantum Chemistry*, edited by E.S. Kryachko and J.L. Calais. Kluwer Academic Publishers, Dordrecht (1994).

B.T. Sutcliffe. J. Mol. Struct. (Theochem) *341*, 217 (1995).

S. Takeno (editor). *Dynamical Problems in Soliton Systems.* Springer-Verlag, Berlin (1985).

V.I. Talanov. JETP Lett. *2*, 141 (1965).

M. Tanaka. J. Phys. Soc. Japan *51*, 2686 (1982).

A.C. Tanner. Chem. Phys. *123*, 241 (1988).

R.A. Tawil and S.R. Langhoff. J. Chem. Phys. *63*, 1572 (1975).

T.R. Taylor (editor). *Optical Solitons. Theory and Experiment.* Cambridge University Press, New York (1992).

A.J. Thakkar. Int. J. Quantum Chem. *23*, 227 (1983).

A.J. Thakkar. J. Chem. Phys. *86*, 5060 (1987).

A.J. Thakkar and W.A. Pedersen. Int. J. Quantum Chem. *24S*, 327 (1990).

A.J. Thakkar and T. Koga. Int. J. Quantum Chem. *26S*, 291 (1992).

A.J. Thakkar, A.M. Simas, and V.H. Smith. Mol. Phys. *41*, 1153 (1980).

A.J. Thakkar, A.M. Simas, and V.H. Smith. Chem. Phys. *63*, 175 (1981).

A.J. Thakkar, A.M. Simas, and V.H. Smith. J. Chem. Phys. *81*, 2953 (1984).

A.J. Thakkar, A.L. Wonfor, and W.A. Pedersen. J. Chem. Phys. *87*, 1212 (1987).

G.F. Thomas. Phys. Lett. *94A*, 265 (1983).

G.F. Thomas, F. Javor, and S.M. Rothstein. J. Chem. Phys. *64*, 1574 (1976).

J.F. Thompson. Appl. Numer. Math. *1*, 3 (1985).

Y. Tourigny. *Product Approximation for Two Nonlinear Klein-Gordon Equations.* Report NA/99, Department of Mathematics, University of Dundee, Dundee, Scotland, U.K. (1987).

Y. Tourigny, E.S. Fraga, J. Ll. Morris, and A.R. Mitchell. *Self Focusing of Solitary Waves in a Three-Dimensional Optical Medium.* Report NA/115, Department of Mathematics, University of Dundee, Dundee (1988).

H.P. Trivedi and E.O. Steinborn. Phys. Rev. *A27*, 670 (1983).

J.H. Van Vleck. Phys. Rev. *31*, 587 (1928).

J.H. Van Vleck. *The Theory of Electric and Magnetic Susceptibilities.* Oxford University Press, Oxford (1932).

V.S. Varma. J. Phys. *A14*, L489 (1981).

J. von Neumann. *Mathematische Grundlagen der Quantenmechanik.* Springer-Verlag, Berlin (1932).

Y.V. Vorobyev. *Method of Moments in Applied Mathematics.* Gordon and Breach Science Publication, New York (1962).

M. Vos and I. McCarthy. Am. J. Phys. *65*, 544 (1997).

F. Wang, D.J. Searles, and E.I. von Nagy-Felsobuki. J. Mol. Struct. *272*, 73 (1992).

J. Wang, B.J. Clark, H. Schmider, and V.H. Smith. Can. J. Chem. *74*, 1187 (1996).

G.N. Watson. *Theory of Bessel Functions.* Cambridge University Press, Cambridge (1966).

J. Wei. J. Math. Phys. *4*, 1337 (1963).

J. Wei and E. Norman. J. Math. Phys. *4*, 575 (1963a).

J. Wei and E. Norman. Proc. Am. Math. Soc. *15*, 327 (1963b).

D.H. Weinstein. Proc. Natl. Acad. Sci. *20*, 529 (1934).

R.J. Weiss, A. Harvey, and W.C. Phillips. Phil. Mag. *17*, 241 (1968).

W.M. Westgate, A.M. Simas, and V.H. Smith. J. Chem. Phys. *83*, 4054 (1985).

W.M. Westgate, A.D. Byrne, V.H. Smith, and A.M. Simas. Can. J. Phys. *64*, 1351 (1986).

W.M. Westgate, R.P. Sagar, A. Farazdel, V.H. Smith, A.M. Simas, and A.J. Thakkar. At. Data Nucl. Data Tables *48*, 213 (1991).

H. Weyl. Math. Annalen *68*, 220 (1910).

H. Weyl. Z. Phys. *48*, 1 (1928).

H. Weyl. *The Theory of Groups and Quantum Mechanics.* Dover, New York (1931).

B.S. Wherrett. *Group Theory for Atoms, Molecules and Solids.* Prentice/Hall International, Englewood Cliffs (1986).

A.B. White. SIAM J. Numer. Anal. *16*, 472 (1979).

H.E. White. Phys. Rev. *38*, 513 (1931).

R.R. Whitehead, A. Watt, G.P. Flessas, and M.A.Nagarajan. J. Phys. *A15*, 1217 (1982).

E.H. Wichmann. J. Math. Phys. *2*, 876 (1961).

E.P. Wigner. Am. J. Phys. *31*, 6 (1963)

B.G. Williams. *Compton Scattering.* McGraw-Hill, New York (1977).

F. Wolf and H.J. Korsch. Phys. Rev. *A37*, 1934 (1988).

D.E. Woon and T.H. Dunning. J. Chem. Phys. *98*, 1358 (1993).

B.G. Wybourne. *Classical Groups for Physicists*. Wiley, New York (1974).

L. Yang, D. Heinemann, and D. Kolb. Chem. Phys. Lett. *192*, 213 (1992).

L. Yang, D. Heinemann, and D. Kolb. Phys. Rev. *A48*, 2700 (1993).

H.C. Yuen and B.M. Lake. Phys. Fluids *18*, 956 (1975).

H.C. Yuen and B.M. Lake. *Nonlinear Wave Concepts Applied to Deep-Water Waves*, in *Solitons in Action*, edited by K. Lonngren and A. Scott. Academic Press, New York (1978).

N.J.A. Zabusky. *A Synergetic Approach to Problems of Nonlinear Dispersive Wave Propagation and Interaction*, in *Proceedings of the Symposium on Nonlinear Partial Differential Equations*, edited by W.F. Ames. Academic Press, New York (1967).

N.J. Zabusky and M.D. Kruskal. Phys. Rev. Lett. *15*, 240 (1965).

V.E. Zakharov. Dissertation, Institute of Nuclear Physics, Siberian Division, USSR Academy of Sciences (1966).

V.E. Zakharov and A.B. Shabat. Soviet Phys. JETP *34*, 62 (1972).

H. Zatzkis. *Boundary Value Problems in Mathematical Physics* in *Fundamental Formulas of Physics*, edited by D.H. Menzel. Dover Publications, Inc., New York (1960a).

H. Zatzkis. *Special Theory of Relativity* in *Fundamental Formulas of Physics*, edited by D.H. Menzel. Dover Publications, Inc., New York (1960b).

12.2 Bibliography

The following chronological compilation lists textbooks covering the general developments of both Quantum Mechanics and Quantum Chemistry, having excluded those on specific topics. The books listed are those available at the University of Alberta Library as well as in the personal collections of the authors and, as a consequence, worthwhile works may have been missed. Furthermore it has not been considered necessary to specify whether the publication date corresponds to the original or to a new edition or printing.

E.U. Condon and P.M. Morse. *Quantum Mechanics*. McGraw-Hill, New York (1929).

J. von Neumann. *Mathematische Grundlagen der Quantenmechanik*. Springer, Berlin (1932).

L. Pauling and E. Bright Wilson. *Introduction to Quantum Mechanics*. McGraw-Hill, New York (1935).

H. Hellmann. *Einführung in die Quantenchemie*. Franz Deuticke, Leipzig (1937).

S. Dushman. *The Elements of Quantum Mechanics*. Wiley, New York (1938).

V. Rojanski. *Introductory Quantum Mechanics.* Blackie and Sons, London (1939).

H. Eyring, J. Walter, and G.E. Kimball. *Quantum Chemistry.* Wiley, New York (1944).

H. Reichenbach. *Philosophic Foundations of Quantum Mechanics.* University of California Press, Berkeley and Los Angeles (1946; second printing).

W. Heisenberg. *The Physical Principles of the Quantum Theory.* Dover, New York (1949).

J.M. Iñiguez Almech. *Mecanica Cuantica.* Memorias de la Academia de Ciencias de Zaragoza, Zaragoza (1949).

E. Persico and G.M. Temmer. *Fundamentals of Quantum Mechanics.* Prentice-Hall, New York (1950).

C.A. Coulson. *Valence.* Oxford University Press, Oxford (1952).

K.S. Pitzer. *Quantum Chemistry.* Prentice-Hall, New York (1953).

F. Mandl. *Quantum Mechanics.* Butterworth. London (1954).

Ph. Pluvinage. *Elements de Mecanique Quantique.* Masson et Cie., Paris (1955).

L.I. Schiff. *Quantum Mechanics.* McGraw-Hill, New York (1955).

W. Heitler. *Elementary Wave Mechanics.* Clarendon Press, Oxford (1956).

H.A. Bethe and E.E. Salpeter. *Quantum Mechanics of One- and Two-Electron Atoms.* Springer, Berlin, and Academic Press, New York (1957).

W. Kauzman. *Quantum Chemistry.* Academic Press, New York (1957).

H.A. Kramers. *The Foundations of Quantum Theory.* North-Holland, Amsterdam (1957).

P.A.M. Dirac. *The Principles of Quantum Mechanics.* Clarendon Press, Oxford (1958).

E.C. Kemble. *The Fundamental Principles of Quantum Mechanics with Elementary Applications.* Dover, New York (1958).

L.D. Landau and F.M. Lifshitz. *Quantum Mechanics.* Pergammon, London (1958).

W.V. Houston. *Principles of Quantum Mechanics.* Dover, New York (1959).

C.W. Sherwin. *Introduction to Quantum Mechanics.* Holt, Rinehart and Winston, New York (1959).

D.R. Bates. *Quantum Theory.* Academic Press, New York (1961).

R.H. Dicke and J.P. Wittke. *Introduction to Quantum Mechanics.* Addison-Wesley, Reading (1960).

J.W. Linnett. *Wave Mechanics and Valency.* Methuen, London (1960).

E. Fermi. *Notes on Quantum Mechanics.* The University of Chicago Press, Chicago (1961).

E. Merzbacher. *Quantum Mechanics.* Wiley, New York (1961).

P. Fong. *Elementary Quantum Mechanics.* Addison-Wesley, Reading (1962).

E. Ikenberry. *Quantum Mechanics.* Oxford University Press, New York (1962).

B. Kursunoglu. *Modern Quantum Theory.* Freeman, San Francisco (1962).

N.F. Mott. *Elements of Wave Mechanics.* Cambridge University Press, Cambridge (1962).

A. Messiah. *Quantum Mechanics.* Halsted, New York (1963).

N.F. Mott and I.N. Sneddon. *Wave Mechanics.* Dover, New York (1963).

D.I. Bloknintsev. *Quantum Mechanics.* Reidel, Dordrecht (1964).

H.A. Kramers. *Quantum Mechanics.* Dover, New York (1964).

A.S. Davydov. *Quantum Mechanics.* Pergamon, Oxford (1965).

R.P. Feynman. *Feynman Lectures on Physics.* Addison-Wesley, Reading (1965).

J.C. Slater. *Quantum Theory of Molecules and Solids.* McGraw-Hill, New York (1965).

J.M. Anderson. *Mathematics for Quantum Chemistry.* Benjamin, New York (1966).

K. Gottfried. *Quantum Mechanics.* Benjamin, New York (1966).

A.A. Sokolov, Y.M. Loskutov, and I.M. Termov. *Quantum Mechanics.* Holt, Rinehart and Winston, New York (1966).

P. Stehle. *Quantum Mechanics.* Holden-Day, San Francisco (1966).

R.L. White. *Basic Quantum Mechanics.* McGraw-Hill, New York (1966).

R.M. Sillito. *Non-Relativistic Quantum Mechanics.* Elsevier, New York (1967).

P.T. Mathews. *Introduction to Quantum Mechanics.* McGraw-Hill, London (1968).

A. Rubinowick. *Quantum Mechanics.* Elsevier, Amsterdam (1968).

D.S. Saxon. *Elementary Quantum Mechanics.* Holden-Day, San Francisco (1968).

J.C. Slater. *Quantum Theory of Matter.* McGraw-Hill, New York (1968).

J.M. Anderson. *Introduction to Quantum Chemistry.* Benjamin, New York (1969).

G. Baym. *Lectures on Quantum Mechanics.* Benjamin, Reading (1969).

J.M. Ziman. *Elements of Advanced Quantum Theory.* Cambridge University Press, Cambridge (1969).

D.B. Beard and G.B. Beard. *Quantum Mechanics with Applications.* Allyn and Bacon, Boston (1970).

J.M. Cassels. *Basic Quantum Mechanics.* McGraw-Hill, London (1970).

D.T. Gillespie. *A Quantum Mechanics Primer.* Wiley, New York (1970).

M. Mizushima. *Quantum Mechanics of Atomic Spectra and Atomic Structures.* Benjamin, New York (1970).

G. Sposito. *An Introduction to Quantum Physics.* Wiley, New York (1970).

J.G. Taylor. *Quantum Mechanics. An Introduction.* George Allen and Unwin, London (1970).

S.R. La Paglia. *Introductory Quantum Chemistry.* Harper & Row, New York (1971).

D. Rapp. *Quantum Mechanics.* Holt, Rinehart and Winston, New York (1971).

D.V. George. *Principles of Quantum Chemistry* . Pergamon, New York (1972).

F.C. Goodrich. *A Primer of Quantum Chemistry.* Wiley-Interscience, New York (1972).

S.P. McGlynn, L.G. Vanquickenborne, M. Kinoshita, and D.G. Carroll. *Introduction to Applied Quantum Chemistry.* Holt, Rinehart and Winston, New York (1972).

R.E. Moss. *Advanced Molecular Quantum Mechanics.* Chapman and Hall, London (1973).

S. Wieder. *The Foundations of Quantum Theory.* Academic Press, New York (1973).

S. Flugge. *Practical Quantum Mechanics.* Springer, New York (1974).

S. Gasiorowicz. *Quantum Physics.* Wiley, New York (1974).

M. Jammer. *The Philosophy of Quantum Mechanics.* Wiley, New York (1974).

H.F. Hameka. *Quantum Theory of the Chemical Bond.* Hafner Press, New York (1975).

C. Cohen-Tannoudji, B. Diu, and F. Laloe. *Quantum Mechanics.* Wiley-Interscience, New York, and Hermann, Paris (1977).

J. Goodisman. *Contemporary Quantum Chemistry.* Plenum, New York (1977).

J.P. Lowe. *Quantum Chemistry.* Academic Press, New York (1978).

A. Bohm. *Quantum Mechanics.* Springer, New York (1979).

W. Pauli. *General Principles of Quantum Mechanics.* Springer, New York (1980).

R. Shankar. *Principles of Quantum Mechanics.* Plenum, New York (1980).

A.T. Fromhold. *Quantum Mechanics for Applied Physics and Engineering.* Dover, New York (1981).

H.F. Hameka. *Quantum Mechanics.* Wiley, New York (1981).

M.W. Hanna. *Quantum Mechanics in Chemistry.* Benjamin, Menlo Park (1981).

A.I.M. Rae. *Quantum Mechanics.* McGraw-Hill, London (1981).

P.W. Atkins. *Molecular Quantum Mechanics.* Oxford University Press, New York (1983).

R.L. Flurry. *Quantum Chemistry. An Introduction.* Prentice-Hall, Englewood-Cliffs (1983).

P.C.W. Davies. *Quantum Mechanics.* Routledge & Kegan Paul, London (1984).

H. Schaefer. *Quantum Chemistry.* Clarendon Press, Oxford (1984).

H.A. Bethe and R.W. Jackiw *Intermediate Quantum Mechanics.* Benjamin-Cummings, Menlo Park (1985).

A. Das and A.C. Milissimos. *Quantum Mechanics. A Modern Introduction.* Gordon and Breach, New York (1986).

T.-Y. Wu. *Quantum Mechanics.* World Scientific, Singapore (1986).

B.H. Bransden and C.J. Jochain. *Introduction to Quantum Mechanics.* Longman, London (1989).

R.E. Christoffersen. *Basic Principles and Techniques of Molecular Quantum Mechanics.* Springer, Berlin (1989).

A. Szabo and N.S. Ostlund. *Modern Quantum Chemistry.* McGraw-Hill, New York (1989).

L.E. Ballentine. *Quantum Mechanics.* Prentice-Hall, Englewood-Cliffs (1990).

A. Galindo and P. Pascual. *Quantum Mechanics.* Springer. Berlin (1990).

F.L. Pilar. *Elementary Quantum Chemistry.* McGraw-Hill, New York (1990).

I.N. Levine. *Quantum Chemistry.* Prentice-Hall, Englewood-Cliffs (1991).

R.K. Prasad. *Quantum Chemistry.* Wiley, New York (1991).

I. Bialynicki-Birula, M. Cieplak, and J. Kaminski. *Theory of Quanta.* Oxford University Press, New York (1992).

C.E. Dykstra. *Quantum Chemistry and Molecular Spectroscopy.* Prentice-Hall, Englewood-Cliffs (1992).

R.L. Liboff. *Introductory Quantum Mechanics.* Addison-Wesley, Reading (1992).

R. McWeeny. *Methods of Molecular Quantum Mechanics.* Academic Press, London (1992).

P.J.E. Peebles. *Quantum Mechanics.* Princeton University Press, Princeton (1992).

D. Park. *Introduction to the Quantum Theory.* McGraw-Hill, New York (1992).

F. Schwabl. *Quantum Mechanics.* Springer, Berlin (1992).

G.P. Schatz and M.A. Ratner. *Quantum Mechanics in Chemistry.* Prentice-Hall, Englewood-Cliffs (1993).

H. Kroemer. *Quantum Mechanics for Engineering, Material Science and Applied Physics.* Prentice-Hall, Englewood-Cliffs (1994).

F.J. Yndurain. *Relativistic Quantum Mechanics and Introduction to Field Theory.* Springer, Berlin (1996).

Appendix. Matrix Notation

Given a square matrix $\mathbf{A}$, with elements A_{pq}, the definitions and notation used in this work are as follows

$\mathbf{A}^{\dagger}$:   *transposed* matrix, with elements A_{qp}.

$\mathbf{A}^{*}$:   *complex conjugate* matrix, with elements A_{pq}^{*} (where the star indicates complex conjugate).

$\mathbf{A}^{-1}$:   *inverse* matrix, such that $\mathbf{A}\mathbf{A}^{-1} = \mathbf{A}^{-1}\mathbf{A} = \mathbf{I}$, where $\mathbf{I}$ is the unit matrix, with elements $I_{pq} = \delta_{pq}$ (δ_{pq} being the Kronecker delta).

$\overline{\mathbf{A}}$:   *Hermitian conjugate* matrix, such that $\overline{\mathbf{A}} = (\mathbf{A}^{*})^{\dagger} = (\mathbf{A}^{\dagger})^{*}$, with elements $\overline{A}_{pq} = A_{qp}^{*}$.

    A matrix $\mathbf{A}$ is said to be an *orthogonal* matrix if $A^{\dagger} = A^{-1}$, in which case $\mathbf{A}\mathbf{A}^{\dagger} = \mathbf{A}^{\dagger}\mathbf{A} = \mathbf{I}$.
    A matrix $\mathbf{A}$ is said to be a *unitary* matrix if $\overline{A} = A^{-1}$, in which case $\mathbf{A}\overline{\mathbf{A}} = \overline{\mathbf{A}}\mathbf{A} = \mathbf{I}$.

Springer
and the
environment

At Springer we firmly believe that an international science publisher has a special obligation to the environment, and our corporate policies consistently reflect this conviction.
We also expect our business partners – paper mills, printers, packaging manufacturers, etc. – to commit themselves to using materials and production processes that do not harm the environment. The paper in this book is made from low- or no-chlorine pulp and is acid free, in conformance with international standards for paper permanency.

Lecture Notes in Chemistry

For information about Vols. 1–32
please contact your bookseller or Springer-Verlag

Editorial Policy

This series aims to report new developments in chemical research and teaching - quickly, informally and at a high level. The type of material considered for publication includes:

1. Preliminary drafts of original papers and monographs

2. Lectures on a new field, or presenting a new angle on a classical field

3. Seminar work-outs

4. Reports of meetings, provided they are
    a) of exceptional interest and
    b) devoted to a single topic.

Texts which are out of print but still in demand may also be considered if they fall within these categories.

The timeliness of a manuscript is more important than its form, which may be unfinished or tentative. Thus, in some instances, proofs may be merely outlined and results presented which have been or will later be published elsewhere. If possible, a subject index should be included. Publication of Lecture Notes is intended as a service to the international chemical community, in that a commercial publisher, Springer-Verlag, can offer a wider distribution to documents which would otherwise have a restricted readership. Once published and copyrighted, they can be documented in the scientific literature.

Manuscripts

Manuscripts should comprise not less than 100 and preferably not more than 500 pages. They are reproduced by a photographic process and therefore must be submitted in camera-ready form according to Springer-Verlag's specifications: technical instructions will be sent on request.

The text area should take care of the page length and width (12.2 x 19.3 cm when you use a 10 point font size, 15.3 x 24.2 cm for a 12 point font size).

Authors receive 50 free copies and are free to use the material in other publications.

Manuscripts should be sent to one of the editors or directly to Springer-Verlag, Heidelberg.